Climate and Energy-Water-Land System Interactions

Technical Report to the U.S. Department of Energy in Support of the National Climate Assessment

DISCLAIMER

This report was prepared as an account of work sponsored by an agency of the United States Government. Neither the United States Government nor any agency thereof, nor Battelle Memorial Institute, nor any of their employees, makes **any warranty, express or implied, or assumes any legal liability or responsibility for the accuracy, completeness, or usefulness of any information, apparatus, product, or process disclosed, or represents that its use would not infringe privately owned rights**. Reference herein to any specific commercial product, process, or service by trade name, trademark, manufacturer, or otherwise does not necessarily constitute or imply its endorsement, recommendation, or favoring by the United States Government or any agency thereof, or Battelle Memorial Institute. The views and opinions of authors expressed herein do not necessarily state or reflect those of the United States Government or any agency thereof.

PACIFIC NORTHWEST NATIONAL LABORATORY
operated by
BATTELLE
for the
UNITED STATES DEPARTMENT OF ENERGY
under Contract DE-AC05-76RL01830

Printed in the United States of America

Available to DOE and DOE contractors from the
Office of Scientific and Technical Information,
P.O. Box 62, Oak Ridge, TN 37831-0062;
ph: (865) 576-8401
fax: (865) 576-5728
email: reports@adonis.osti.gov

Available to the public from the National Technical Information Service
5301 Shawnee Rd., Alexandria, VA 22312
ph: (800) 553-NTIS (6847)
email: orders@ntis.gov <http://www.ntis.gov/about/form.aspx>
Online ordering: http://www.ntis.gov

PNNL-21185

Climate and Energy-Water-Land System Interactions

Technical Report to the U.S. Department of Energy in Support of the National Climate Assessment

Coordinating Lead Authors: Richard Skaggs, Pacific Northwest National Laboratory
Kathy Hibbard, Pacific Northwest National Laboratory

Lead Authors: Peter Frumhoff, Union of Concerned Scientists
Thomas Lowry, Sandia National Laboratories
Richard Middleton, Los Alamos National Laboratory
Ron Pate, Sandia National Laboratories
Vince Tidwell, Sandia National Laboratories

Contributing Authors: Jeffrey Arnold, USDA-Agricultural Research Service
Kristen Averyt, University of Colorado
Anthony Janetos, Pacific Northwest National Laboratory
Cesar Izaurralde, Pacific Northwest National Laboratory
Jennie Rice, Pacific Northwest National Laboratory
Steve Rose, Electric Power Research Institute

DOE Program Manager: Bob Vallario, DOE-Office of Science

Prepared for
the U.S. Department of Energy
under Contract DE-AC05-76RL01830

Pacific Northwest National Laboratory
Richland, Washington 99352

Executive Summary

This technical input report on climate and energy-water-land (EWL) system interactions has been prepared for the U.S. Department of Energy in support of the U.S. National Climate Assessment (NCA). Prepared on an accelerated schedule to fit the NCA's timeline, it provides a summary of existing information and understanding of this broad topic.

This report provides a framework to characterize and understand the important elements of climate and EWL system interactions. It identifies many of the important issues, discusses our understanding of those issues, and identifies the research needs to address the priority scientific challenges and gaps in our understanding. Much of the discussion is organized around two discrete case studies with the broad themes of (1) extreme events and (2) regional differences. These case studies help demonstrate unique ways in which energy-water-land interactions can occur and be influenced by climate. In addition, a series of "illustrations" portray representative decision-making considerations relevant to climate-EWL interfaces. Key findings from the report are summarized below, according to the report section in which they are found.

Characterization of Climate and Energy-Water-Land System Interactions

- Population growth and economic and social development are major drivers of the demand for energy, land, and water resources within the interdependent climate and EWL system. A major challenge will be to manage, and optimize where possible, competing economic and environmental objectives and priorities within resource budget constraints and impact risks of climate variability and change.

- The interdependencies of climate and the EWL system can be characterized by the three bilateral interfaces of energy-water, energy-land, and land-water. Each bilateral interface consists of linkages representing the supplies, end-use demands, and associated functional relationships between the two.

- Much of our current understanding of climate impacts on the complex interdependencies of the EWL system is derived from limited observations of bilateral interface responses to climate variability. The concept of EWL interfaces can help identify the relative degree of risks and vulnerabilities to the effects of climate variability and change.

Energy-Water-Land Interfaces: Resource Interdependencies and Interactions with Climate

- Focusing on sector-to-sector interfaces alone does not adequately capture the complexity and importance of the EWL system. The many bilateral interfaces form a dynamic set of interacting processes linked through a complex network of feedbacks.

- Competition for water is the most straightforward conflict linking energy, water, and land (e.g., simultaneous demand for thermoelectric generation, irrigation, environmental flows).

- Extreme climate events such as drought and associated heat waves have important impacts on the EWL interfaces. Impacts are seen as changes in cropping and grazing and accompanying wildfire damage. These changes tend to reinforce and intensify individual impacts on land and water resources (e.g., reduced cropping raises feed prices, which changes grazing patterns, which in turn affects vegetation density and thus wild fire vulnerability). To a lesser extent, these changes feed back through water and land use to impact energy demand and production.

- U.S. regions differ in their (a) current climate, (b) projected climate change, (c) energy mix (e.g., solar/wind availability, coal), (d) energy supply and demand, (e) water availability/regularity and water sources (e.g., rain vs. snow fed), and (f) the availability and quality of land. Each region will be differentially impacted by climate change and each region will have to adapt or mitigate using different strategies. The manner in which adaptation strategies and associated institutions evolve has significant implications for energy-water-land dynamics.

Risk, Uncertainty, and Vulnerability Associated with Climate Impacts on Energy-Water-Land Interfaces.

- Risk, uncertainty, and vulnerability at the EWL interfaces are minimally reported in the literature, and where they are documented, they are usually case-specific. However, risk, uncertainty, and vulnerability are generally found to have the following four characteristics.

  – They are broader in scope.

  – They can be amplified or attenuated across sectors.

  – They have altered temporal and spatial dynamics.

  – They manifest during extreme (low-likelihood, high-consequence) events.

  These characteristics are fundamental to understanding how risk, uncertainty, and vulnerability relate to each characteristic across sectors, and for developing solutions and strategies that may reduce their impact or influence

- Awareness of the four characteristics of risk, uncertainty, and vulnerability can enable monitoring and assessment mechanisms to anticipate and avoid or correct unintended consequences, and that more effective emergency planning and response can be put in place by understanding the sources of those consequences.

- Perhaps the biggest gains will come by increasing our understanding of human response and behavior to climate change and decisions concerning climate change, identifying the trigger points where low-probability events within one sector can become high-consequence events in other sectors, and in identifying and understanding the amplification, attenuation, and feedback mechanisms that create unintended and unanticipated consequences.

Climate Mitigation and Adaptation at the Energy-Water-Land Interfaces

- Many mitigation and adaptation options tie directly into one of the EWL sectors, and are therefore tied into the EWL interfaces. Understanding the EWL nexus is therefore central to the effective design, selection, implementation, and monitoring of adaptation and mitigation strategies.

- Almost all mitigation options lie within either the energy or land sectors. Mitigation reduces or sequesters emissions arising from the supply and demand for energy and land (e.g., substituting renewable technologies for fossil fuel generation; preventing deforestation). As such, mitigation options are affected by EWL relationships.

- Adaptation options designed to reduce vulnerability to climate impacts in one EWL sector affect, and are affected by, EWL linkages. Some adaptation measures reduce demands on EWL endowments (e.g., water-use efficiency), while others may increase them (e.g., desalinization).

- Understanding how mitigation and adaptation strategies relate to the EWL interfaces facilitates not only the evaluation of the net impact of individual mitigation or adaptation measures, but also the compound effects of concurrent implementation, either intentionally or as an outcome of the uncoordinated actions of independent parties.

- Some sector-specific mitigation and adaptation measures have the potential to provide synergistic "win-win" opportunities to enhance climate mitigation or adaptation objectives across one or more other sectors in the nexus. Other measures may have negative impacts on mitigation or adaptation potential in other sectors.

Research Needs Associated with Climate Impacts on Energy-Water-Land Interfaces

- A major complication in understanding and responding to climate changes is that they are often characterized by multiple interactions, feedbacks, and tradeoffs among different human activities and environmental processes.

- Simulating and understanding the interactions and feedbacks among climate and the EWL system requires not only accurate representations of each individual sector, but also a detailed understanding of the scale-dependent interactions among them.

- Addressing the climate-EWL related questions that regional decision makers are asking will require the development of models capable of evaluating different adaptation strategies, testing different mitigation options, and accounting for the tradeoffs, co-benefits, and uncertainties associated with these actions or combinations of actions—such as how technology cost, performance, and availability will impact results.

- Research needed to substantially increase our understanding of the interactions and feedbacks among energy, water, land, and climate include the following:

  - Meaningful analyses of the EWL interfaces will require a new class of models, measurements, and observations that are consistent with global climate and socio-economic constraints, and capable of resolving regional human decision-making and natural processes in a manner that captures the full range of relevant interactions and feedbacks.

  - New models, observing systems, or, even modifications of existing frameworks will require new strategies for understanding and quantifying uncertainty.

  - Evaluation strategies that account not only for individual model performance, but properties of coupled systems will require robust metrics for benchmarking model performance and data systems.

  - For regions, or areas that are unable to provide the current data required (e.g., recent transportation infrastructure, pricing policies, demography, environmental information), new methods into parsing sparse data or extracting information from pre-existing data will be required.

  - In the context of climate, support for new research and modeling capabilities that account for potential future environmental constraints (e.g., availability of water), economic limitations (e.g., existing infrastructure) and scenario development will be needed to inform decision making processes for the deployment of future energy transitions.

- New tools that provide stakeholders with information on near-term consequences, as well as long-term implications of options that might be considered to mitigate or adapt to climate change will be needed.

- Accounting for natural boundaries, such as watershed, energy utility or geo-political zones will need to be incorporated into existing gridded calculations and observations.

Acronyms and Abbreviations

AA	amplification and attenuation
CAFO	concentrated animal feeding operation
CBM	coalbed methane
CCS	carbon capture and storage
CHP	combined heat and power
CONUS	conterminous United States
DET	demand-endowment-technology
DOE	U.S. Department of Energy
EIA	U.S. Energy Information Administration
EWL	energy-water-land
ERCOT	Electricity Reliability Council of Texas
GCRA	Global Change Research Act
GDP	gross domestic product
GHG	greenhouse gas
IPCC	Intergovernmental Panel on Climate Change
LEAP	Long-range Energy Alternatives Planning (model)
MAF	million acre-feet
NCA	National Climate Assessment
NREL	National Renewable Energy Laboratory
PDSI	Palmer Drought Severity Index
PV	photovoltaic
RCP	Representative Concentration Pathway
RO	reverse osmosis
R&D	research and development
REMI	Regional Economic Models Incorporated
RPS	Renewable Portfolio Standard
SARF	Social Amplification of Risk Framework
SARP	Societal Applications Research Program
USGCRP	U.S. Global Change Research Program
WCI	Western Climate Initiative
WECC	Western Electricity Coordinating Council
WEAP	Water Evaluation and Planning (model)
WWT	wastewater treatment

Contents

Figures

Tables

1.0   Introduction

The Global Change Research Act (GCRA) of 1990 requires the U.S. Global Change Research Program (USGCRP) to conduct National Climate Assessments (NCAs), no less than every four years. These assessments are meant to analyze the effects of global climate change on the natural environment, agriculture, energy production and use, land and water resources, transportation, human health and welfare, human social systems, and biological diversity. Further, they are meant to evaluate current trends in global change and project major trends for the next 25–100 years. These assessments also are intended to evaluate our progress toward reducing risk, vulnerability, and impacts, as well as the implications of alternative adaptation and mitigation options, and provide a sustained process for informing an integrated research program.

Previous NCAs have evaluated climate impacts by sector (i.e., water resources, energy supply and use, transportation, agriculture, ecosystems, human health, society). The third NCA, scheduled for completion in 2013, will address for the first time the impacts of climate in the context of "sectoral cross-cuts" (i.e., assessment of climate impacts on the interdependencies of two or more sectors). One such sectoral cross-cut is how energy, water, and land use compete against and constrain one another in the context of climate change and management decisions.

As a part of the current NCA effort, member agencies of the USGCRP and external researchers are providing technical inputs on the topics of their choosing that will be relevant to one or more of the chapters in the 2013 NCA report. As one of the responsible agencies, and consistent with its long-term energy and research mission, the U.S. Department of Energy (DOE) is providing technical input on the cross-sectoral topic of climate change and energy, water, and land system interactions. This document serves as DOE's technical input report on this topic.

Past national assessments have examined climate impacts on the energy, water, and land sectors individually, with limited emphasis on cross-sectoral impacts. In fact, while the research community has become increasingly interested in interactions across these sectors, information and understanding of combined climate impacts and cross-sectoral interdependencies remain limited. For example, Dale et al. (2011) found no studies on the combined effects of climate on land use and energy.

Consequently, this technical input will, for the first time in the context of the NCA, focus on climate impacts on the combined interdependencies of energy, land, and water as they apply to the overall support of natural ecosystems, infrastructure, and human socio-economic activities. Because research in these areas is still in the early stages, much of the information presented is preliminary, identifying potential issues and the need for additional observation, analysis, and modeling.

1.1   Schedule

This report began with a request from DOE to develop a technical input report describing the current state of knowledge, both empirical and using models, of the interactions among energy, water, land and the climate system. An author team was quickly assembled, and the authors benefited from a scoping workshop in July 2011, which produced a working outline. After a period of reviewing the literature and identifying more detailed topical outlines, an expert workshop was held in November 2011 to provide an opportunity for broader input from the scientific community. The author team's subsequent drafts of chapters in this report have been reviewed internally by the team of 15 authors, and a mail review was conducted in February of 2012 prior to reconciliation of comments and delivery to DOE and the NCA on March 1, 2012. For the mail review the document was sent to 15 reviewers, and 5 sets of comments were received.

1.2 Overview

This report highlights much of the limited information on and understanding of climate and energy-water-land (EWL) system interactions in the context of issues, potential impacts, and long-term research needs. The report begins with a detailed characterization of the climate-EWL nexus and associated issues in terms of the interfaces between the three interdependent energy, water, and land resource sectors. A conceptual model is presented that defines the EWL nexus in terms of resource supply and demand linkages. Using this model, the report briefly describes the paired bilateral interfaces of energy-water, energy-land, and land-water, as well as the integrated three-part system of energy-water-land interfaces. It also includes examples of supply-demand linkages and processes for selected human and ecosystem support applications. The report then explores how individual bilateral interfaces interact in response to climate. Next, the report addresses risk, uncertainty, and vulnerability in the context of sector interfaces. Mitigation and adaptation decision-making vulnerabilities, opportunities, and coordination are then discussed in light of their EWL relationships. Finally, long-term research needs are discussed in the context of challenges and opportunities with regard to data completeness and accuracy; requirements for integrated modeling including energy, water, and land systems; and identified risks, vulnerabilities, and uncertainties.

This report does not provide a detailed description of current and future impacts of climate on the U.S., but builds on reports such as *Global Climate Change Impacts in the United States*, June 2009, in which the USGCRP describes current and future impacts of climate change on various U.S. regions and sectors. Observed trends include rising temperatures, increasing heavy rainfall, changes in the amounts and timing of river flows, and many others. Impacts are shown to vary by U.S. region and analyzed relative to seven specific sectors: water, energy, transportation, agriculture, ecosystems, health, and society.

1.3 Approach

This report was developed by an author team under DOE oversight, with significant input from participants in the July and November 2011 workshops. Data, methods, and tools varied in depth and completeness, being derived from available source materials, primarily in the published literature. The report content, assessment findings, and levels of confidence reflect consensus among the report authors, considering comments from selected external reviewers.

The NCA provided broad guidance for the technical input reports, including identifying eight topics that are priorities for the 2013 report: (1) risk-based framing; (2) confidence characterization and communication; (3) documentation, information quality, and traceability; (4) engagement, communications, and evaluation; (5) adaptation and mitigation; (6) international context; (7) scenarios; and (8) sustained assessment (http://www.globalchange.gov/images/NCA/nca-priority-topics-guidance.pdf).

The objectives for this effort were to (1) provide a framework to characterize and understand the important elements of climate and energy-water-land system interactions, (2) identify the important issues and our level of understanding of those issues, and (3) present a long-term research program to address the priority gaps in scientific understanding.

Given the challenge of comprehensively exploring the full breadth of the climate-energy-water-land interface in a relatively short report, this report focuses on a detailed investigation of two examples, constructed as case studies with the broad themes of (1) extreme events and (2) regional differences. These themes were selected because they address key issues facing resource managers and policy makers, and because they help demonstrate some unique ways in which the individual bilateral interfaces can interact across the broader spectrum of the energy-water-land triad and respond to different climate scenarios. In this context, specific instances of climate change and/or variability are used to illustrate potential implications of future climate change. In addition to the case studies, several "illustrations" or representative examples are provided to illustrate the consequences of decision-making processes on bilateral interfaces. The illustrations describe specific examples of climate-EWL interactions and associated decision-making issues (these are found in sections 2 and 3, set apart as blue text boxes).

1.4   Why the Energy-Water-Land Nexus?

A core goal of this report is to provide decision makers with an understanding of how the convergence of supply and demand issues related to energy, water, and land resources in a changing climate are central to informing planning and policy choices about climate mitigation and adaptation. Mitigation and adaptation measures are typically aimed at one or more of the interfaces. Analyzing mitigation and adaptation through the lens of the energy-water-land interface facilitates not only the evaluation of the net impact of individual mitigation or adaptation measures, but also the compound effects when they are implemented together, either intentionally, or as is more likely, as an outcome of uncoordinated actions of independent parties. These compound effects may not always have positive synergies, and ignoring or failing to identify potentially negative interactions, runs the risk of undermining planning and policy goals (Moser 2012).

1.5   References

Dale VH, RA Efroymson, and KL Kline. 2011. "The land use – climate change – energy nexus." *Landscape Ecology* (26):755-773.

Moser SC. 2012. "Adaptation, mitigation, and their disharmonious discontents: an essay." *Climatic Change* 111(2):165-175, DOI: 10.1007/s10584-012-0398-4.

2.0   Energy-Water-Land Interfaces: Resource Interdependencies and Interactions with Climate

2.1   Introduction

2.1.1   Focus and Approach

This section focuses on the assessment of climate variability and change effects through the interactions of climate with the supply and use of energy, land, and water (EWL). This approach recognizes that the EWL resource sectors are dynamically linked to one another in a complex interdependent system that supports and interacts with human socio-economic activities, infrastructure, and ecosystems. As depicted in Figure 2.1, this system also operates within the context of climate variability and change, making it even more complex. The terminology used to describe this interdependent system is the climate- EWL nexus. Within this nexus, bilateral supply and demand linkages exist across the interfaces between each resource sector pair. Each resource sector will also influence, and be influenced by, climate variability and change, as shown in Figure 2.1.

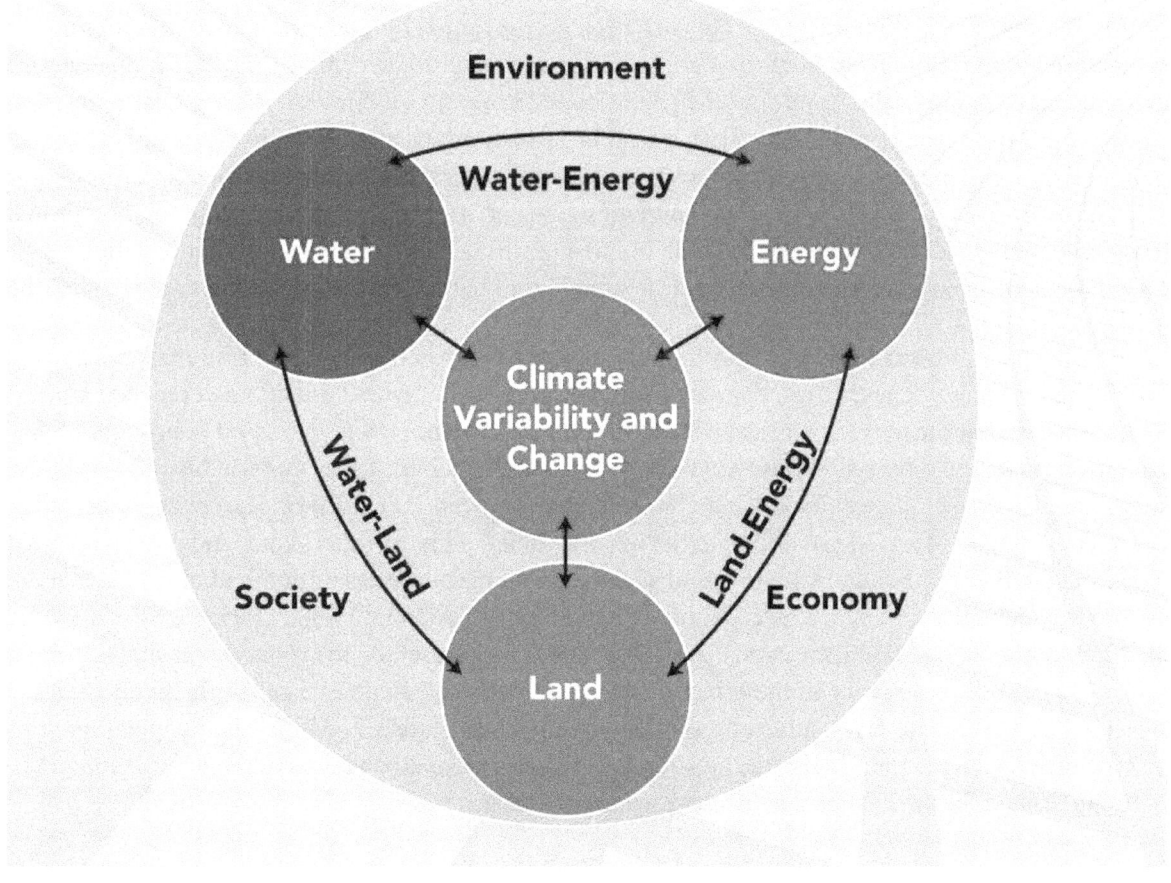

Figure 2.1. The climate-EWL nexus.

Figure 2.2 elaborates on Figure 2.1 by illustrating some of the features and attributes within each resource sector. It also depicts the bilateral linkages into supply-demand linked pairs between sector interfaces. The linked pairs are expressed as energy for water, water for energy, energy for land, land for energy, water for land, and land for water. The color-coded legend on the right side of the figure gives representative examples of application functions, facilities, and infrastructure for each supply-demand interface link.

The remainder of this section further describes and characterizes the framework for assessing climate change effects through the lens of the climate-EWL nexus, including several example illustrations. This lens will be applied in later sections of the report using more detailed case study scenarios and the consideration of mitigation and adaptation strategies to manage risks of climate variability and change impacts across the EWL interfaces (IPCC 2007; CCSP 2008; Karl et al. 2009; NAS 2010a).

2.1.2   Framing Climate Assessment Through Energy-Water-Land Interactions

Human population growth accompanied by increasing levels of technological development over time have led to greatly increased use of the Earth's land, water, energy, mineral, and biological resources for meeting human socio-economic needs (Dale et al. 2011). Human-induced changes globally have become so significant over the past two centuries that they are now seen as marking a new Anthropocene epoch (Crutzen and Stroemer 2000; Zalasiewicz et al. 2010; Ellis et al. 2010; Zalasiewicz et al. 2011). The period from 1800 to 2000 saw global population expand from less than 1 billion to 6.5 billion, with a projected increase to 9 billion by 2050. This has been accompanied by the development of cities and a general demographic shift away from rural communities to major urban centers, which is a characteristic land use change feature of the Anthropocene (Zalasiewicz et al. 2011). Similarly, the global terrestrial biosphere went from being predominantly in a wild or semi-natural state with only very minor agricultural and settlement use in 1700, to now having only a quarter wild and less that 20% semi-natural. The majority land use is now in agriculture and cities (Ellis et al. 2010). This human population growth and socio-economic advancement was enabled by the industrial and agricultural revolutions fueled by the dramatic increase in consumption of energy, water, and other natural resources (Zalasiewicz et al. 2011).

The expansion of human socio-economic activities and associated demand on resources has caused major changes in land use and cover, degraded soil and water quantity and quality, and reduced land carbon sequestration capacity. It has also altered carbon flux balances, and increased atmospheric greenhouse gas (GHG) concentrations, which are now more than a third above pre-industrial levels. Such changes are affecting climate and inducing climate change impacts (IPCC 2007; Karl et al. 2009); adding stress to ecosystems (MEA 2005); affecting the quantity and quality of water, land, and energy resources; and threatening the well-being, resiliency, and security of human societies (Gordon et al. 2003; Marland et al. 2003; Feddema et al. 2005; Lotze-Campen et al. 2005; MEA 2005; Pielke 2005; CCSP 2008; Wise et al. 2009; Smith et al. 2010; Dale et al. 2011; DSB 2011; Ebinger et al. 2011; Ferguson et al. 2011; NASA 2011). The convergence of these factors and their interdependencies can be effectively represented by the climate-EWL interactions of Figure 2.1 and Figure 2.2, with human population growth and associated socio-economic activities, resource use practices, and behaviors recognized as major underlying drivers for the demands on energy, water, and land.

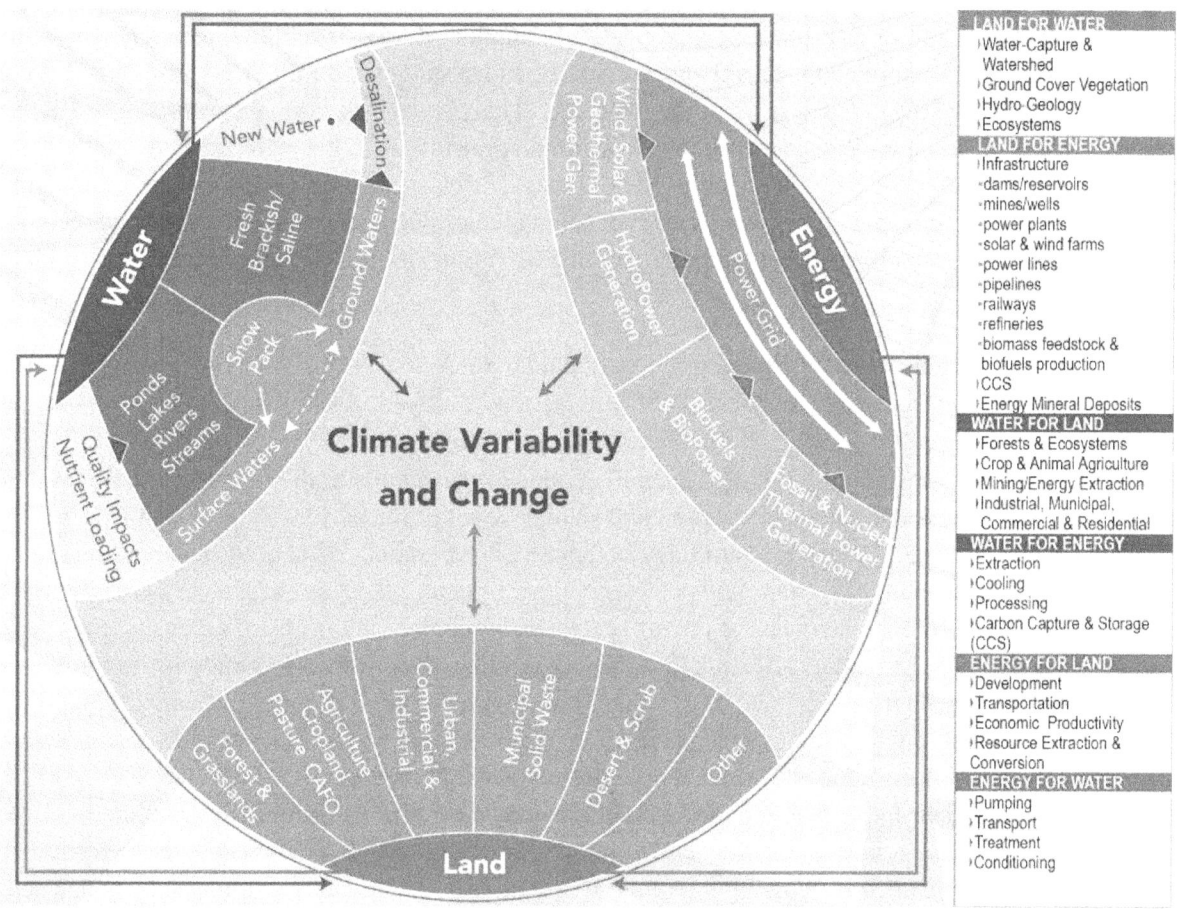

The following image contains the following legend text:

LAND FOR WATER
›Water-Capture & Watershed
›Ground Cover Vegetation
›Hydro-Geology
›Ecosystems

LAND FOR ENERGY
›Infrastructure
  -dams/reservoirs
  -mines/wells
  -power plants
  -solar & wind farms
  -power lines
  -pipelines
  -railways
  -refineries
  -biomass feedstock & biofuels production
›CCS
›Energy Mineral Deposits

WATER FOR LAND
›Forests & Ecosystems
›Crop & Animal Agriculture
›Mining/Energy Extraction
›Industrial, Municipal, Commercial & Residential

WATER FOR ENERGY
›Extraction
›Cooling
›Processing
›Carbon Capture & Storage (CCS)

ENERGY FOR LAND
›Development
›Transportation
›Economic Productivity
›Resource Extraction & Conversion

ENERGY FOR WATER
›Pumping
›Transport
›Treatment
›Conditioning

Figure 2.2. Illustration of the climate-EWL nexus showing linkages and interactions among the three resource sectors with climate variability and change.

2.1.3   Resource Interface Linkages and Climate Interdependencies

Energy, water, and land resources and associated support ecosystems constitute the foundation on which all human societies rely for their existence, productive development, security, and well-being. All three resource sectors are highly coupled to one another through supply-demand relationships that support both human socioeconomic activities and the ecosystems on which societies rely for critical services (Daily et al. 1997; MEA 2005; Smith et al. 2010; Dale et al. 2011). The bilateral interface linkages are the building blocks of the EWL nexus. Assessments of interface interactions and possible impacts of climate variability and change have been investigated to a limited extent by focusing on only one sector at a time (Marland et al. 2003; Feddema et al. 2005; Lotze-Campen et al. 2005; Pielke 2005; Oki et al. 2006; IPCC 2008; CCSP 2008; Brekke et al. 2009; Stone 2009; BOR 2011; Ebinger et al. 2011; FIPCCWDI 2011; Vine 2011), or by targeting one or two of the sector interfaces and linkages (Izaurralde et al. 2003; DOE 2006; Pate et al. 2007; Atlantic Council 2011; Dale et al. 2011; Ferguson et al. 2011; GLC 2011; McMahon et al. 2011; Outka 2011; Wise et al. 2009; Warren 2011). A more complete assessment requires considering all three resource sector interfaces together as an integrated system.

The bilateral EWL interfaces can each be characterized by supply and demand linkages and mutual influences, as shown in Figure 2.2 and Figure 2.3. The linkages can be described in terms of a resource *demand* to meet a natural or man-made need, a resource supply *endowment*, and the related *technologies* (including natural and man-made mechanisms, processes, and systems) and influences that come into play on both the supply and the demand sides of the interface. This *demand-endowment-technology* (DET) interface linkage description is discussed in more detail in Appendix A. This is introduced simply as a notional construct that recognizes and denotes the obvious existence of complex and dynamic functional relationships among atmospheric, geophysical, geochemical, biological, and human engineered technical and socio-economic systems operating within the climate-EWL nexus at different temporal and spatial scales.

The interdependencies among the energy, water, and land resource sectors and climate for a given situation will involve a complex mix of nonlinear interactions and feedbacks of varying intensity and timing delays (Izaurralde et al. 2003; Marland et al. 2003; Feddema et al. 2005; Oki et al. 2006; Ferguson et al. 2011; Mahowald 2011). These will depend on the details of the interacting EWL system configuration under consideration, the location and spatial extent of the various system elements, and the temporal scales involved. The interactions have both natural and human-influenced features because they are clearly part of both natural systems and socio-economic systems. They may also vary greatly by geographic region. The characteristics of the processes, technologies, and infrastructures that may specifically come into play under different situations for both the supply and utilization of resources may also make a difference in the magnitude of the interactions with climate. Technologies are also expected to change over time through technological advancements at rates that may differ from the operational lifetimes of existing infrastructures using technologies of varying vintages. Similarly, the rates of change of technologies and their implementation infrastructures may each differ significantly from the frequency and duration of climate variations and the periods of climate change. This can be expected to result in more complex and dynamically evolving interactions and impacts of climate variations and change across the climate-EWL nexus.

An example is the energy-water-land interface situation represented by a large urban center's demands for water and electric power. It is assumed in this example that power will be generated by a combination of hydroelectric and thermoelectric power plants operating within the local watershed, with the thermoelectric generation fueled by a combination of energy minerals and biomass. Also assumed is that biomass is purposely grown for energy production in the region where irrigation is supplied by a combination of surface and ground water. Water for the many municipal, industrial, commercial, and residential uses within the urban center is assumed to be provided by a water supply infrastructure consisting of a combination of surface and ground fresh water sources coupled to treatment facilities that feed a transport and distribution network of pipelines, pumping stations, and storage tanks. Both hydropower and water-cooled thermoelectric power, as well as the urban demand for water, rely on water supplies that depend on natural hydrogeological cycle processes (Oki et al. 2006) that include precipitation and watershed capture and transfer via surface and subsurface flows (including snowpack storage and timed release). Soil moisture and evapotranspiration also play a role (Oki et al. 2006; Bales et al. 2011), particularly for production of the biomass feedstock. The man-made technologies, processes, and infrastructures involved will include a combination of (with high-level interface linkages indicated in parentheses):

- dams and reservoirs (land use for both water and energy)
- water turbine and generator facilities (water use for energy)

- surface and ground water control and transport infrastructure (land use for water; energy use for water)
- irrigation infrastructure for biomass production (land use for water & energy; water use for energy; energy use for water
- biomass production, harvesting, processing, transport, and storage infrastructure (land use for energy; water use for land; water use for energy)
- thermal power plant generation and cooling technologies and facilities (land use for energy; water use for energy)
- power plant energy mineral fuel (coal, natural gas, nuclear) extraction, processing, transport, and storage infrastructure (land use for energy; energy use for energy; water use for energy)
- electric power transmission and distribution infrastructure (land use for energy)
- urban center water supply production, treatment, and distribution infrastructure (land use for water; energy use for water)
- urban center end-use technologies, infrastructures, and practices that govern the demand and efficiencies of electric power and water use (land use for water & energy; energy use for water; water use for energy)

Potential climate impacts on the supply-demand interface linkages for the above example include the following:

- Changes in the quantity and timing of hydrographic cycle flows result in changes in reservoir storage capacity and in hydropower generation. Such changes can also impact downstream uses, including water availability for biomass irrigation, for thermoelectric power plant cooling, and for other urban center uses;

- Shifts in heating and cooling degree-days can alter timing for peak power demands and affect loads on the grid and transmission systems. Higher air temperatures, and water temperatures for cooling intakes, also mean less efficient electricity generation and transmission. This can be problematic during heat waves when air conditioning loads are high. Periods of prolonged drought and/or high daily temperatures can impact biomass production by degrading growth and yields and/or requiring irrigation that increases water demand;

- Competing water demands in arid or over-allocated water systems under climate stress (drought and/or heat wave) translate into less water available for thermoelectric power generation (resulting in potential reductions in power output or loss of power), and/or less water for urban center supplies (resulting in higher water costs, use limitations, increased pumping and stress on ground water supplies), and/or less water for irrigation of biomass (resulting in less biomass for energy production or more use of pumped groundwater that puts more demands on power and more stress on ground water supplies);

- Extreme climatic events (floods, storms, etc.) can damage installed infrastructure for both energy and water systems, affecting their operations and resulting in delays and capital expenditures for repairs and restoration of service.

Appendix B provides a high-level inventory matrix showing examples of bilateral interface linkages for energy-water, energy-land, and water-land arranged by columns that are associated with a list of selected applications arranged by row. The column headings also include the notional linkage functions shown in Figure 2.3 and in Table A.1. As the details of the interface linkage functions and their interactions between sectors and with climate are better understood and can be expressed or modeled quantitatively with greater spatial and temporal resolution and precision, the more informative the analysis and assessment can potentially be. To the extent this can be accomplished, it may provide greater insight and help reduce uncertainty for decision support. Even in the absence of more accurate quantitative modeling and forecasting, methodologies of risk assessment and risk-based mitigation and adaptation strategy development can be applied (NAS 2010; Ebinger et al. 2011). However, the critical importance of using more integrated assessment approaches that take into consideration cross-sector and multi-regional interactions is becoming more recognized and beginning to be applied (GLC 2011; Dale et al. 2011; Dale et al. in-press; Warren 2011).

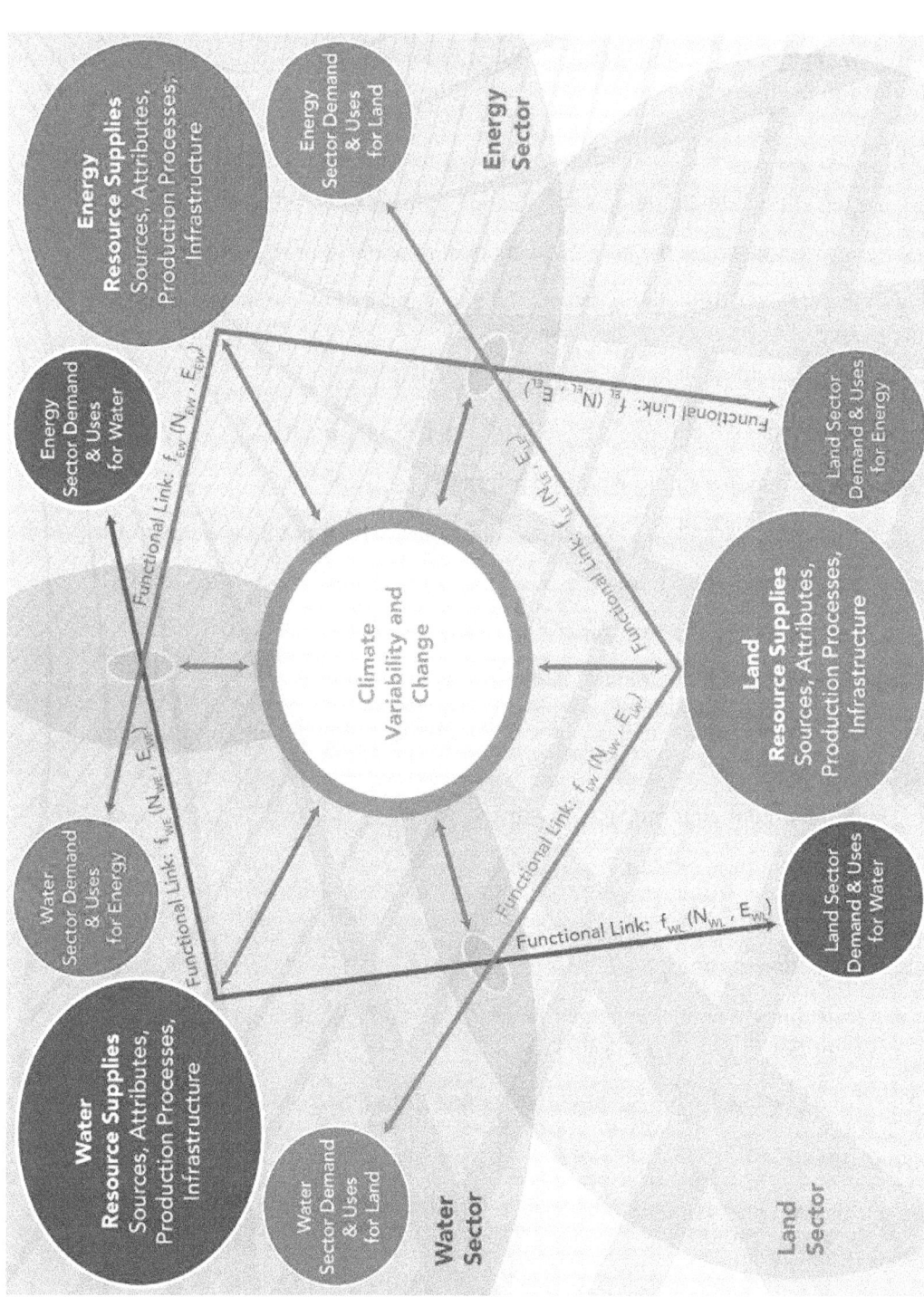

Figure 2.3. Interdependent interface linkages for integrated climate-EWL nexus, with functional linkages described in more detail in Appendix A. Functional notation shown on the links represents natural and man-made processes and systems that come into play in supplying a resource from one sector to meet an end-use demand in another sector. Bilateral interaction arrows (green) represent the influence of climate on resources and linkage processes and systems, and vice versa. Resource supply and demand attributes, linkages, and related processes and infrastructure are distinguished by color coding: brown for land, red for energy, and blue for water.

2.7

2.2   Integrated Framework and Approach to EWL Interdependencies

2.2.1   Issues and Trends in the U.S.at the Climate-EWL Nexus

There are a number of emerging U.S. trends that are relevant to the climate-EWL nexus. A sampling of such trends is provided in the list below.

Energy (AEO 2012):

- Domestic energy use has declined with the recent economic downturn
  - Projected slower growth rate over the next 25 years than previously projected (AEO 2012)
  - Combination of increased energy efficiency and the expectation of an extended period required for economic recovery
  - Overall energy consumption per capita is projected to decline
- Prices for crude oil in 2011 stayed in range $85 - $110… up to $145 (in 2010 $) in 2035
- Significant uncertainty surrounding future world oil prices
- Diesel prices higher than gasoline and expected to maintain that…stronger global demand for diesel
- Natural gas prices stay below $5/MCF until after 2023
- Coal prices projected to increase 1.4% per year (from $ 1.76 per MMBTU in 2010)
- Electricity prices decline, reflecting low natural gas prices, but higher than 2011 due to rising coal $$
- Biofuel consumption has been growing and is expected to continue growth ...but projection not realistic because of E10 barrier for EtOH
- Consumption of liquid fuels continues modest growth
- Natural gas consumption continues growth
- Coal consumption increases 0.3% per year
- Renewable fuels consumption grow 2.8% per year
- Energy use per capita declines at annual average rate of 0.5%..total population grows 25% by 2035, but total energy consumption grows 10% in same period
- Slow growth in electricity demand + low/competitive natural gas prices = declining coal demand

Water (Kenny et al. 2009):

- Total U.S. water use in 2005 was 410,000 million gal per day … 85% fresh (349,000 million gal per day)
- Fresh water withdrawals for thermoelectric-power generation and irrigation, the two largest uses of water, have stabilized or decreased since 1980.

- Withdrawals for public-supply and domestic uses have increased steadily since estimates began. Withdrawals for irrigation in 2005 were 128 Bgal/d, about 8 percent less than in 2000 and approximately equal to estimates of irrigation water use in 1970.

- Thermoelectric-power withdrawals accounted for 49 percent of total water use, 41 percent of total freshwater withdrawals, and 53 percent of fresh surface-water withdrawals for all categories.

- In 2005, irrigation withdrawals accounted for 37 percent of all freshwater withdrawals and 62 percent of all freshwater withdrawals excluding thermoelectric withdrawals.

- Irrigation withdrawals represented 31 percent of all water withdrawals, and 37 percent of all freshwater withdrawals

- Irrigated acreage increased from 25 million acres in 1950 to 58 million acres in 1980, then remained fairly constant before increasing in 2000 and 2005 to more than 60 million acres.

- The number of acres irrigated using sprinkler and micro-irrigation systems has continued to increase and in 2005 accounted for 56 percent of the total irrigated acreage. Withdrawals for irrigation of crops and other lands totaled 128,000 Mgal/d and were the second-largest category of water use.

- The Renewable Fuels Standard (NAS 2011b), established by federal energy legislation (EISA 2007), has influenced the demand for biofuels. Other renewable energy sources such as wind and solar also of interest and seeing growth.

- Application of biofuels and other renewable energy technologies can likewise influence water and land use (NAS 2007; Groom et al. 2008; Kocoloski et al. 2009; Scott et al. 2010; Tyner et al. 2010; Carter 2011, Kobos et al. 2011; Scott et al. 2011).

- Biofuels and biopower are expected to displace fossil fuel use and reduce carbon emissions. However, energy, land, and water are needed to grow the biomass for these energy sources, and success depends heavily on climate, weather, and geographic location.

- Increasing competition for fresh water is driving the use of "new" water through treating and using non-fresh water sources (DOE 2006), which increases energy demand but also makes good use of underused or marginal water and land resources.

Climate Change (DSB 2011; Karl et al. 2009):

- Trends seen with climate based on observations over multiple decades

- Increasing land and sea surface temperatures

- Changing ocean temperature

- Changing ocean chemistry (acidity and salinity, with impacts on ecosystems and circulation)

- Declining mass of Greenland and Antarctic ice sheets

- Declining glaciers and snow cover

- Decreasing and thinning Arctic sea ice

- More frequent and longer droughts

- Increased frequency of heavy precipitation events, flooding, and landslides

- Increased cyclone intensity

- Rising sea level

U.S. Population (2010): 308,746,000:

- Population growth and its associated socio-economic development and resource use practices continue to be major factors driving demand for energy, water, and land use change (Ellis et al. 2010; Smith et al. 2010; Zalasiewicz et al. 2010; Dale et al. 2011; Zalasiewicz et al. 2011)

- Each resource sector within the climate-EWL nexus will have sustainable supply limitations, but are being increasingly competed for by multiple users.

- Each resource will be subject to the supply and demand dynamics and constraints that link across sectors (Berndes 2002; Lotze-Campen et al. 2005; Reilly et al. 2007; Reilly et al. 2008; Rosegrant et al. 2009; Scott et al. 2009; Stillwell et al. 2009; Vine 2009; Scott et al. 2010; Smith et al. 2010; Cooley 2011; Dale et al. 2011)

- Allocating limited resources among competing uses requires making tradeoff decisions with implications for climate variability and change

- Managing competing economic, social, and environmental goals and priorities within budget constraints, while evaluating risks and strategies for mitigation or adaptation to possible impacts of climate variability and change, presents a major challenge for decision makers (Berndes 2002; Falkenmark 2003; Falkenmark et al. 2005; Lotze-Campen et al. 2005; Brekke et al. 2009; Smith et al. 2010; Cooley et al. 2011; Dale et al. 2011; Vine 2011)

2.2.2   Bilateral Interfaces in the Context of Climate

2.2.2.1   Effects of Energy on Water Use

There are a wide range of energy demands for water resources including water for: operating and cooling thermoelectric power plants; geothermal and concentrating solar power (CSP) plants; dam and reservoir storage for hydropower generation; fracking fluid for shale gas production and drilling fluid for oil and gas wells; mining and processing coal and uranium; extracting and processing oil shale and tar sands; and for growing and processing biomass (Cooley et al. 2011; McMahon et al. 2011). Implementation of carbon capture and storage (CCS) on power plants to capture CO_2 from flue gas would require additional water and power (Kobos et al. 2011; Williams et al. 2011). Some of these demands are described in detail below.

The majority of thermoelectric power generation (fueled by coal, natural gas, biomass, or uranium) uses water for steam turbine operation that requires boiler water and cooling water. This technology generated over 80 percent of U.S. electricity in 2009. Water for thermal power plant is withdrawn from both fresh and brackish or saline water sources. Thermoelectric power generation uses the largest fraction (estimated at 201 billion gallons per day [BGD], or 41 percent) of all fresh water withdrawals in the U.S. in 2005 (Kenny et al. 2009). Nearly all (99 percent) of the water withdrawals for U.S. thermoelectric power are from surface sources, and of these, 28 percent were saline water withdrawals (Kenney et al. 2009). A more recent assessment for 2008 estimated thermoelectric power plant freshwater withdrawals of between 60 to 170 BGD, with a consumption of 2.8 to 5.9 BGD, which is 4.7 to 5.9 percent (Averyt et al. 2011). The type of power plant and its cooling system, and the geographic location of the plant and source of water, will determine the water use intensity and the impact on water supplies. Power plants in the Southwest with relatively scarce surface water often withdraw from ground water aquifers that may already suffer from overdraft. Western power plants also use more recirculating closed cycle cooling than Eastern plants with once-through cooling. Water withdrawals per unit of power produced are far lower with closed cycle, but water consumption is higher (Averyt et al. 2011; Macknick et al. 2011). A nuclear plant with once-through cooling has a median withdrawal of 44,000 gallons per MWh and median consumption of about 270 gal per MWh. With closed cycle cooling the median withdrawal is 1,100 gal per MWh and the consumption is about 670 gal per MWh (Macknick et al. 2011). Dry cooling can reduce cooling water demand to zero, while hybrid cooling water demand falls between dry and closed cycle (McMahon et al. 2011). Using coal as a fuel demands additional water (5 to 70 gal per MWh) for mining and washing to remove impurities, and scrubbers to remove sulfur dioxide from the emissions. Similarly, uranium needs additional water (45 to 150 gal per MWh) for fuel processing (McMahon et al. 2011). In the case of once-through cooling, water at elevated temperature may end up be returned to the environment and impact the downstream ecosystem.

Conventional hydropower production is based on the use of a dam on a river that creates a storage reservoir, redirects river flow, and provides power generation, flood control, recreation, and general surface water management. The U.S. has over 79,000 dams of all sizes, and the world has about 50,000 dams that are considered to be large with a height of 15 meters or higher (McMahon et al. 2011). Hydropower causes little or no air pollution, but can negatively affect aquatic habitats and ecosystems. It is currently the largest source of renewable energy in the world, has increased by 50% since 1990, and in 2008 generated about 16.3 percent of global electricity production. In 2009 the U.S. generated 8.6 percent (282 TWh) of total annual domestic production. Hydropower is subject to climate effects that reduce the quantity and timing of precipitation, snow pack, and water flows. Reservoirs also lose water to evaporation and are subject to silt build-up from upstream erosion, which can impact their operation and life (McMahon et al. 2011). Average evaporative water loss for hydropower is 4,500 gal per MWh (McMahon et al. 2011). Evaporation will increase with higher temperatures and lower humidity, and erosion and silting will increase with more severe rain events and the loss of watershed tree cover that accompanies forest fires and/or dead trees from insect damage that is promoted by drier, hotter temperatures.

Biomass as feedstock for either biopower or biofuels requires water to grow and process the biomass into fuel. With ample precipitation, biomass may not need irrigation. However, lack of adequate soil moisture during the growth process can reduce the biomass yield, and supplemental irrigation may be needed in dry periods to assure reliable biomass growth and harvest. Excessive water and fertilizer application can also result in nutrient-rich water runoff that can contaminate adjacent surface or ground water bodies, which can adversely impact the health and services of local ecosystems. Hydrologic/land-energy feedback with climate can exist and will depend on local conditions, whether irrigation is used, and whether the water source is surface or subsurface (Ferguson et al. 2011). Using biomass from forest thinning can help watersheds by reducing fuel for wildfires and improving water capture and ground water recharge in forested regions. The use of agricultural crop wastes and perennial lignocellulosic[1] energy crops grown in regions with adequate precipitation can reduce the water-use intensity and water quality impacts of biofuels.

---

[1] *Lignocellulous* is the primary structural component of all plants and is a renewable non-food material.

Extremes in temperature can also adversely impact biomass growth. Prolonged drought coupled with prolonged high temperatures during the summer can damage the crop. Such conditions can also reduce surface water supplies and drive up demand for power for air conditioning and for pumping ground water, which puts greater load demand on power plants and drives greater demand for water. Extreme weather events can damage or destroy the crop and associated infrastructure.

Carbon capture and storage (CCS) demands water to strip CO_2 from flue gas and power to process concentrated liquefied CO_2 (Williams et al. 2011). A recent analysis estimated the parasitic power loss at a coal-fired power plant for CCS operation to be 20 percent of power plant capacity, with an increase in water demand of 43 percent (Kobos et al. 2011). These added burdens with CCS offsets some of the benefit. Although a low carbon emitting technology, CSP generation can also have high water use in the range of 750 – 920 gal per MWh. Geothermal can also be high in water use, depending on the technologies used, with median water use intensity ranging from 0 to about 4,780 gal per MWh (Macknick et al. 2011;McMahon et al. 2011).

Wind and solar photovoltaics (PV) renewable energy systems are low carbon emitting and also low water use. The power generated is intermittent and is subject to variations in local wind and solar resource conditions. Solar also only operates during daylight hours. The intermittency presents a challenge to electric utilities who must compensate by having adequate conventional back-up generation capacity to cover times of low wind and solar resource and peaks in electricity demand. Although expanded use of wind and solar can theoretically reduce demand for water for power production, until improved technology is developed, the amount of intermittent renewable energy generation capacity that can be added to the utility grid system will be limited by stability issues and back-up power requirements. Climate variations and change can impact these systems in several ways. Extreme weather events, such as high winds and floods, can damage infrastructure. Temperature extremes can

Using treated water from municipal wastewater treatment (WWT) facilities as cooling and processing water for energy and power production or irrigation water for energy crops can productively re-use the water and make dual use of the energy already expended to operate the WWT facility. Similarly, developing and implementing cooling and other industrial water use technologies that would enable the use non-treated brackish or saline water can avoid the demand for additional energy for water treatment.

Mining and well drilling for energy and fertilizer mineral extraction demands water, as does, the downstream processing or refining of these materials into fuels. These processes can also contaminate water, which leads to additional energy demands and costs to treat the water for proper disposal or re-use. Water is a significant byproduct associated with petroleum, natural gas, and coal bed methane exploration and production. This "produced water," may contain a variety of contaminants. If produced water is not appropriately managed or treated, these contaminants may present a human health and environmental risk (Newell et al. 2006; GAO 2012). Fracking of shale gas requires large quantities of water that becomes contaminated as waste water. The fracking process can also cause earthquakes and the intrusion of fracking fluid into fresh water aquifers, causing subsurface contamination of drinking water supplies. Mining also requires water and can cause water contamination with adverse impact on ecosystems. Treatment of produced water and other wastewater from mining and energy extraction operations requires additional energy and water.

Additional insights into the effects of energy on water and associated decision making considerations are presented in Illustrations 2.1 and 2.2.

Illustration 2.1. Climate-Related Decision Making at the Nexus:
Thermoelectric Power and Water Availability

Local water resources are stressed when there is an imbalance between demands relative to surface and groundwater supplies. Across the U.S., water supplies in the Western U.S. tend to be under more pressure than those in the East (Figure 2.4a). This is not surprising given the relatively plentiful surface water supplies east of the Mississippi, compared with the arid regions of the West. However, power plants account for over 70% of water withdrawals in the East, such that water demands for thermoelectric cooling are contributing to water stress (Figure 2.4b). Based on the ratio of water demand to water supply, power plants are the major drivers of water stress in 44 basins across the United States.

| (a) Water supply stress in the U.S. | (b) Water supply stress by power plants. |

Figure 2.4. (a) Water supplies across the U.S. are stressed from multiple demands on the system.
(b) In some places, stress on water supplies is driven by power plant cooling water requirements. (Source: Averyt et al., 2011)

Climate change has the potential to significantly alter the national picture of water stress. Changes in the hydrologic cycle are already being observed across the United States, and climate change is expected to continue to alter the amount of water available—meaning more water on average in some places, and significantly less in others (Karl et al. 2009). Droughts and floods will continue to be a part of the U.S. climatological landscape, but in the long term (post-2100) it is expected that droughts in the southern half of the U.S. will intensify (IPCC 2007; Karl et al. 2009). In the short term (2050), it is unclear how variability in the climate system will be affected by climate change.

Consideration in Decision Making: See Illustration 2.4.

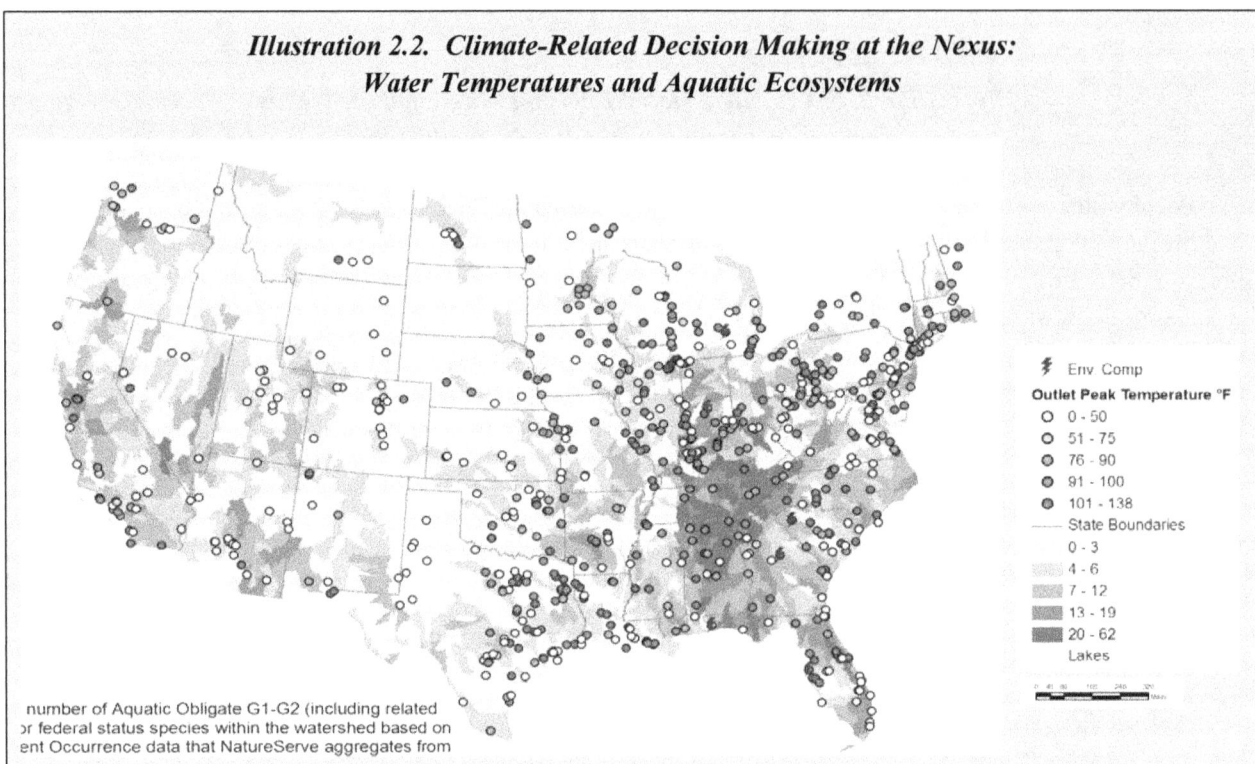

number of Aquatic Obligate G1-G2 (including related
or federal status species within the watershed based on
ent Occurrence data that NatureServe aggregates from

Figure 2.5. Power plant effluent water temperatures and aquatic habitats in the conterminous U.S.

Figure 2.5 shows the compiled datasets for maximum temperatures of intake and outflow water for power plant cooling units from EIA Form 923, "Power Plant Operations Report," in 2008. Temperature discharge data are overlain on total counts of aquatic-obligate species (NatureServe 2011). *Aquatic obligate* species are those ranked G1-G2 by NatureServe, or that have federal status under the U.S. Endangered Species Act. *Aquatic obligate* is defined as any species known to spend all or part of its lifecycle living in the water, based on NatureServe's habitat data, or on taxonomic group in the absence of habitat data. These species include fish, crayfish, mussels, and dragonflies, for example.

Increasing pressure on water supplies to accommodate multiple demands is expected in regions where long-term trends in average water availability are diminished by climate change. However, for many purposes the quantity of water available is also a function of the quality of that water. Both power plants and ecosystems require quantities of water within an optimal temperature range.

Thermoelectric power plants that use once-through cooling technologies withdraw water from a nearby lake, stream, or river. Water that is not consumed is generally returned to the source, but at a temperature that is, on average, 17°F warmer than the original intake water temperature (Madden 2010).

Peak summer temperatures for return flows from more than 350 power plants across the country exceeded 90°F—which is the thermal discharge limit set by 14 states (Averyt 2011; EPA 2011). High water temperatures can imperil fish and other aquatic species (Hester and Doyle 2011). Many trout species cannot be exposed to temperatures exceeding 80°F for more than a few minutes. Several bass species can tolerate temperatures up to 100°F, but some power plants discharge water at temperatures up to 110°F (Averyt 2011). High effluent temperatures have resulted in power plant closures or curtailment of operations. Duke Energy had to cut power generation at its G.G. Allen and Riverbend coal-fired plants, as the temperature of discharged cooling water exceeded thermal limits. The closures in the Carolinas resulted in power blackouts (Beshears 2007).

However, elevated temperatures on the intake part of the once-through cooling process can also be problematic. Not only does a higher water temperature increase effluent temperature that much more, but warm water decreases the efficiency of the cooling process, and reduces electricity output relative to the amount of water withdrawn. On several occasions (2007, 2010, 2011), the temperature of the Tennessee River rose above 90°F. Since the Browns Ferry Nuclear Plant is a once-through facility that draws water from the Tennessee, the temperature of the intake water ensured that the effluent would exceed the 90°F thermal limit, forcing curtailment of operations at the power plant (Nuclear Regulatory Commission 2011; 2010; 2007). It is estimated that the 2010 shutdown alone cost ratepayers over $50 million (Smith 2011).

Consideration in Decision Making:

Water temperatures are largely a function of ambient air temperature and relative water quantities. Increasing air temperatures, in combination with regional changes in water quantities, will impact water temperatures in natural aquatic systems. In general, elevated water temperatures in streams, lakes, and rivers can be expected as a result of climate change (Karl et al 2009).

Issues at the nexus between energy and water in the context of climate change, water temperature, and ecosystem impacts may be expected east of the Mississippi, as most U.S. power plants using freshwater resources for once-through cooling are in the East. Already, as a consequence of deleterious thermal impacts on aquatic ecosystems, some power plants have switched from once-through cooling technologies to evaporative cooling. However, there are tradeoffs. Although water withdrawals are minimal relative to once-through facilities, and thermal impacts to aquatic habitats are avoided, water consumption can be three times greater when cooling towers are employed (Macknick et al. 2011). Also, the plant would still not be immune to changes in the efficiency of electricity generation resulting from elevated water temperatures at the intake.

A few power plants in Florida and South Carolina have constructed cooling towers downstream to cool effluent water before it is reintroduced into the environment (Averyt et al. 2011). Although this presumably minimizes impacts to aquatic ecosystems, water consumption would increase substantially depending on the frequency at which these cooling towers are operated.

2.2.2.2   Effects of Water on Energy Use

Water demands on energy resources include pumping, treatment, processing, and conditioning water for various public, private, commercial, industrial, mining, and agricultural end-uses (DOE 2006; Atlantic Council 2011; McMahon et al. 2011). Pumping requires energy for the extraction of groundwater and for transport of both surface and groundwater over horizontal distances that require overcoming head loss due to pipeline or channel flow and the traversing of uphill changes in elevation. Energy is also required for the treatment (filtering, demineralization, disinfection) of potable supplies, for the desalination of brackish or saline waters, to for the treatment of various wastewaters (municipal, agricultural, industrial), and to condition water for various end uses (heating, cooling, and further processing for industrial purposes). About 52,000 community water systems use energy to treat and deliver drinking water to over

290 million Americans (GAO 2011). Residential consumers use it for a variety of purposes, including for drinking; bathing; preparing food; washing clothes and dishes; and flushing toilets, which can represent the single largest use of water inside the home. Energy is needed to accomplish many of these activities, which may include filtering and softening water and heating it for use in certain appliances. This accounts for 12.5 percent of a typical household's energy use (GAO 2011). In addition to residential water users, commercial, industrial, and institutional customers use energy to produce hot water and steam for heating buildings, to cool water for air conditioning buildings, and to generate hot water needed to manufacture or process food and materials.

After water is used by customers, energy is needed to collect and treat wastewater and discharge the treated effluent to an appropriate body of water. It may also be further treated, using more energy, to allow for productive re-use. Wastewater service is provided to more than 220 million Americans by about 15,000 municipal wastewater treatment facilities (GAO 2011). Water-related energy use in California consumes about 20 percent of the state's electricity, 30 percent of its natural gas, and 88 billion gallons of diesel fuel every year Atlantic Council 2011, while the U.S. used over 123 million MWh of electricity in 2002 to meet its water service demand (to supply public, domestic, commercial, industrial, mining and energy minerals, livestock, and irrigation; also treatment of public and private wastewater), corresponding to about 425 kWh per person annually (McMahon et al. 2011). These processes consume about 4% of the total electricity generated in the U.S. and the water and wastewater industry is the third largest electricity consumer nationally (McMahon et al. 2011). The energy use intensity of providing water services is highly dependent on type of water source (ground or surface), water quality, source location relative to the processor and the customer, and the additional processing and technology used by the end-use customer (GAO 2011; McMahon et al. 2011).

As demands increase for limited fresh water supplies in the future, the treatment and re-use of non-fresh water and the desalination of brackish or saline water is expected to grow. These processes are relatively energy-intensive, with the result that the demand for energy to provide water services can also be expected to increase (DOE 2006; Atlantic Council 2011; McMahon et al. 2011). Growth in yields of produced water from oil, gas, and coal bed methane, and in waste fracking water from shale gas production is included in the mix of non-fresh water and wastewater sources needing treatment.

In addition to direct demands for energy from the water sector, another important decision making consideration is the implications of this intense energy demand on GHG emissions. This topic is explored further in Illustration 2.3.

Illustration 2.3. Climate-Related Decision Making at the Nexus:
Water in the West: Where Adapting to Diminishing Water Resources Contributes to GHG Emissions

Energy is required to pump, treat, distribute, and use potable water, and to treat and discharge wastewater. The energy intensity of water, or the energy used to provide a unit of water (a gallon, acre-foot, etc.), depends on the source and quality of the raw water, and the type of use. For example, pumping raw water over long distances or over mountain ranges can use a large amount of electricity; California's State Water Project and Arizona's Central Arizona Project are well-known examples. Many cities in the West rely on high quality water that flows to city treatment plants by gravity, requiring very little energy to pump, treat, and distribute the water to customers; increasing urban water supplies will, in many cases, require cities to pump water over greater distances or from deeper aquifers.

The energy intensity of water will vary depending on the source (i.e., surface or groundwater) and the quality of the water. Cities that rely on surface water fed from snowmelt in the Rocky Mountains (e.g., Denver) generally require only moderate amounts of energy to treat and distribute water. For example, the energy intensity of treating and distributing water in Denver in 2007 was 232 kWh/AF (Western Resource Advocates 2008). Colorado Springs has also relied primarily on gravity-fed water supplies from the Rocky Mountains. To expand its supplies, Colorado Springs recently began construction on the Southern Delivery System, a project that will pump water from Pueblo Reservoir to Colorado Springs, requiring an estimated 4,631 kWh/AF (not including treatment or distribution) (U.S. Bureau of Reclamation 2008).

In many parts of the West, where water demands already exceed supplies, there is already a need to import water between watersheds and across state lines, and tap additional groundwater resources (e.g., Colorado Water Conservation Board 2010, Texas Water Development Board 2012). These different projects require varied quantities of energy (Table 2.1).

Existing and Proposed Water Supply Projects

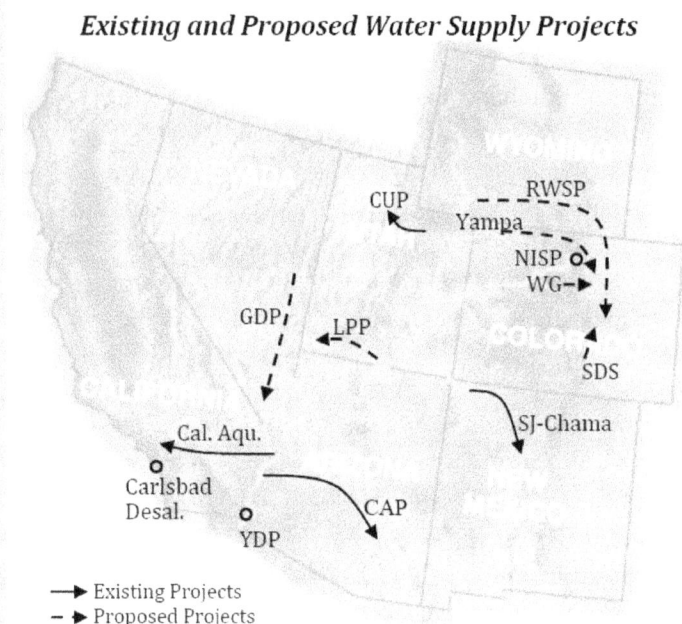

Abbreviation	Project Name
Cal. Aqu.	California Aqueduct
CAP	Central Arizona Project
Carlsbad Desal.	Carlsbad Desalination Plant
CUP	Central Utah Project
GDP	Groundwater Development Project
LPP	Lake Powell Pipeline
NISP	Northern Integrated Supply Project
RWSP	Regional Watershed Supply Project
SDS	Southern Delivery System
SJ-Chama	San Juan-Chama Project
WG	Windy Gap Firming Project
Yampa	Yampa Pumpback Project
YDP	Yuma Desalting Project

⟶ Existing Projects
- ▸ Proposed Projects

Energy Intensity of the West's Water Supplies

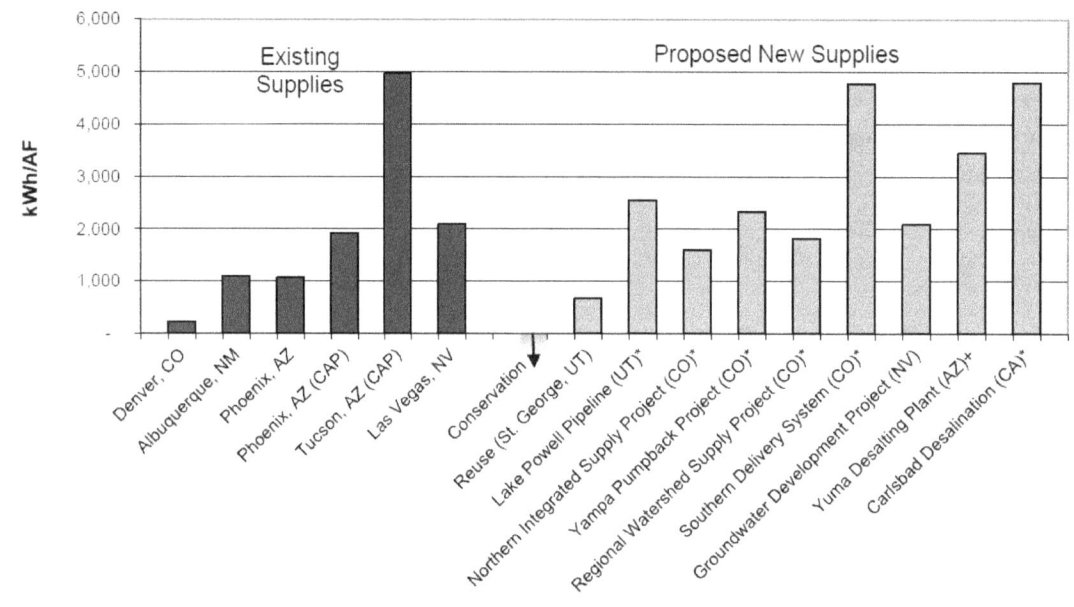

Figure 2.6. The energy intensity of many proposed projects exceeds the energy intensity of existing supplies. Notes: *Figures include an estimated 150 kWh/AF for treatment and/or distribution. ⁺The Yuma Desalting Plant includes the energy used on site and the energy used to pump water to participating utilities in Arizona, Nevada, and California, as its operation is designed to increase water supplied to cities in those states. Colorado Springs' Southern Delivery System and the Carlsbad Desalination Plant are now under construction. The upper map only includes the Colorado River system.

Table 2.1. Energy Use for Water Delivery at Selected Project Sites in the Southwest

Project (State)	Water Delivery (AF/yr)	Net Energy Use (MWh/yr)	Sources
Lake Powell Pipeline (UT)	86,000	307,020	Volume data: Utah Department of Water Resources. 2011. Draft Lake Powell Pipeline Study Water Needs Assessment, prepared by MWH. March 2011. Energy Intensity data: Utah Board of Water Resources. 2008. Lake Powell Hydroelectric System Notice of Intent to File an Application for Original License (Volume 1).
Northern Integrated Supply Project (CO)	40,000	33,980 - 57,980	U.S. Army Corps of Engineers. 2008. Northern Integrated Supply Project Draft Environmental Impact Statement, table 4-15. April 2008.
Yampa Pumpback Project (CO)	300,000	595,680	Northern Water Conservancy District. 2006. Multi-basin Water Supply Investigation.
Regional Watershed Supply Project (CO)	250,000	199,000 - 496,000	Volume data: Wyco Power and Water, Inc. 2011. Application for Preliminary Permit, Regional Watershed Supply Project. Filed with the Federal Energy Regulatory Commission August 31, 2011. Energy Intensity data: Derived from Preliminary Permit Application; analysis summarized in Western Resource Advocates. 2011. Motion to Intervene in Opposition and Comments on the Preliminary Permit Application for the Regional Watershed Supply Project, Project No. P-14263-000. Filed with FERC on December 15, 2011.
Southern Delivery System (CO)	52,900	246,038	U.S. Bureau of Reclamation. 2008. Southern Delivery System Final Environmental Impact Statement. December 2008.
Groundwater Development Project (NV)	217,655	383,073	Volume data: Southern Nevada Water Authority. 2011. Southern Nevada Water Authority Clark, Lincoln, and White Pine Counties Groundwater Development Project. Conceptual Plan of Development. Prepared for the U.S. BLM, March 2011. Energy data: Derived from power capacity needs and an assumed operating rate of 85% (data provided in SNWA, 2011).
Carlsbad Desalination Plant	56,000	260,680	City of Carlsbad, CA. 2005. Precise Development Plan and Desalination Plant Project Environmental Impact Report, p. 4.2-19.

Consideration in Decision Making:

The energy impacts of adapting to changing water supplies are an important consideration; some may help cities both adapt to and mitigate climate change, while others help cities adapt, but increase greenhouse gas emissions. The energy requirements necessary for adapting to climate-driven changes in water supply is an example of how decision making about climate adaptation can conflict with efforts to mitigate greenhouse gas emissions

The impact of climate change on the energy used for water depends on a host of factors. A shift in availability of water supplies may prompt water providers to develop more water supplies that require pumping from greater depths (groundwater) or conveyance over longer distances. Water providers may need to rely increasingly on lower quality supplies that require more extensive treatment, such as tapping more saline supplies that require reverse osmosis (RO). The energy intensity of RO depends on the salinity of the water treated; for example, in its demonstration run in 2007, the Yuma Desalting Plant used approximately 1,451 kWh/AF to treat brackish water (salinity of 2,539 mg/L, reduced to 252 mg/L).[2] In addition to changing water availability, climate change may affect the timing and magnitude of runoff. For many water utilities, existing storage facilities may adequately accommodate variable runoff regimes. Some utilities, however, may require additional storage. If "new" storage includes aquifer recharge (and subsequent recovery), it may lead to additional energy demands. Finally, wastewater treatment plants often discharge treated wastewater into streams; this depends on adequate stream flows to ensure that discharges do not exceed stream temperature or water quality standards. Reduced stream flows or elevated stream temperatures may drive wastewater treatment plants to increase treatment standards, increasing the energy intensity of treatment.

Managing the impacts of diminished and changing water supplies can be informed by current adaptation strategies. New water supply projects such as the Southern Delivery System (Table 2.1) may increase and diversify a water utility's water supply portfolio, but also increase total energy demands. Alternative options include water conservation, increasing use of recycled water, and developing flexible leasing arrangements between cities and farmers. Each of these options has different benefits. Water conservation can both reduce total water demands and save energy, particularly if conservation efforts focus on reducing the use of hot water and/or energy-intensive water conveyance or pumping systems. Recycled wastewater is typically drought resistant; depending on the level of treatment required to provide recycled water, it may have additional energy demands. UV disinfection, for example, is energy intensive. However, the energy used to treat and distribute recycled water may be less than the energy required for new water supply projects. Under traditional agricultural-urban leasing agreements, cities pay farmers to temporarily fallow a portion of irrigated agricultural land and transfer water to cities; these agreements may enable cities to mitigate the impacts of more extreme droughts without increasing the need for energy intensive new infrastructure projects..

2.2.2.3   Effects of Energy on Land Use

Energy demands on land resources include the use of land to grow biomass and the infrastructure to harvest, process, transport, store, and distribute the material for use in the production of biopower or biofuel. Land with forest cover can also provide trimmings and forest industry woody wastes as another source for bioenergy or biofuels feedstock (Cook et al. 2011). Similarly, agricultural land and operations (farms, ranches, dairies, feedlots) can provide biomass residue as additional bioenergy feedstock (DOE 2011).

Also needed to support energy is land for roads, railways, and facilities. Land is also needed for the siting of dams and reservoirs for hydropower, thermal and geothermal power plants, concentrator solar thermal power plants, wind and solar PV generation, and the associated electrical transmission and distribution lines and switching infrastructure needed to control and transport power to end-users.

Land will also contain energy minerals and other deposits to be mined or drilled for extraction. Land would then be needed to site the mining and drilling facilities, and the extracted material processing and transport infrastructure. In the case of solid mined materials like coal and uranium, transport would need

---

[2] Data: 3,819 MWh of electricity were used during the operation of the plant, and 2,632 AF were treated. Source: U.S. Bureau of Reclamation, 2008. Yuma Desalting Plant, Demonstration Run Report.

roads and railways. For natural gas and petroleum, it would require pipelines and tanker trucks on railways or roadways. Land for processing and disposal of waste streams or byproducts like spent nuclear material, produced water, fracking fluid, and captured CO_2 for sequestration. This would include facilities and infrastructure needed for these functions and operations. Depending on where and how land use and land management practices associated with energy are done, ecosystems may either benefit or be adversely impacted. (As an example, Illustration 2.4 discusses impacts of renewable energy development on terrestrial ecosystems). High density development of energy infrastructure resulting in the fracturing and splitting of habitats and the possible release of air emissions and ground and water discharges would risk having adverse effects, especially on sensitive ecosystems. Sustainable forest and agricultural land use management practices in conjunction with ecosystems that include bioenergy feedstock production can increase net carbon sequestration rates, contribute to long-term carbon storage, and contribute to the production of fuels with lower GHG emissions (Koshel et al. 2008; Koshel et al. 2010).

Climate variations and change can impact the hydrological cycle and the growth and productivity of biomass (Oki et al.). Other energy related land uses that involve facilities and infrastructure can also be impacted by extreme weather events that can cause damage and disruption of service. The land use associated with energy can also interact with climate and have influence through release of greenhouse gases and aerosols, change of ground cover and albedo, and reduction or enhancement of carbon capture and sequestration in vegetation and soil.

Illustration 2.4. Climate-Related Decision Making at the Nexus:
Renewable Energy Development and Terrestrial Ecosystems

Energy development can directly affect wildlife through changing land use. Construction and decommissioning can include the direct mortality of wildlife, as well as indirect effects through destruction and modification of habitat. Operation and maintenance of energy facilities can have deleterious effects on ecosystems as they can create habitat fragmentation and barriers to gene flow (Lovich and Ennan 2011). Different technologies have different land footprints (Figure 2.7). Assuming future trends towards increasing renewables, up to 200,000 km^2 of land could be impacted by new energy development by 2030 (McDonald et al. 2009).

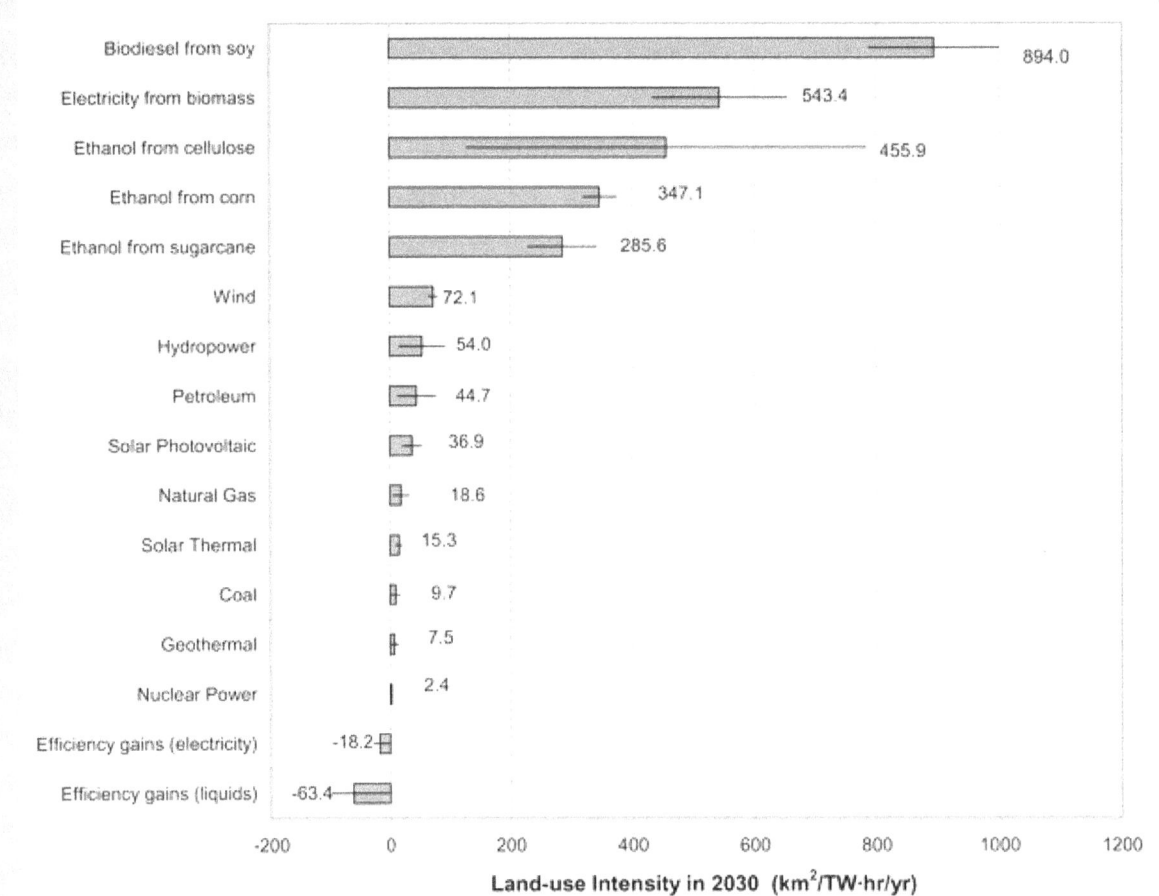

Figure 2.7. Land-use intensity estimated for 2030 associated with energy development.
(Source: McDonald et al. 2009)

The greatest potential for development of renewable energy technologies is in the Southwestern United States (WGA 2009; Figure 2.8). The West is also home to vast expanses of land that host migratory species important to the ecosystem and to Western culture. However, development of renewable energy, particularly wind and solar, can create problems for wildlife if appropriate siting is not considered. This issue could be exacerbated in the future as efforts to mitigate GHG emissions by increasing investment in renewable energy collide with climate driven changes in habitat and migratory patterns.

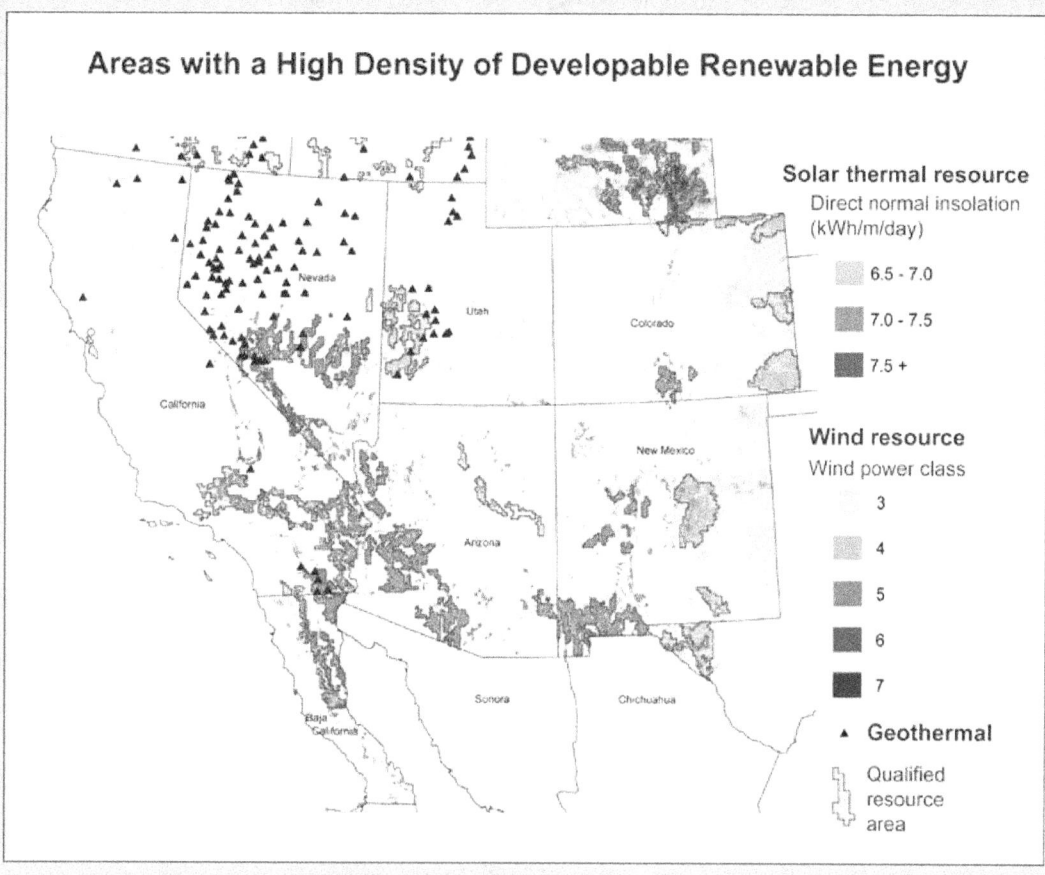

Figure 2.8. Renewable energy potential in the Southwestern U.S. Figure from WGA and DOE data (see Southwest Climate Change Network 2012)

Consideration in Decision Making:

Over the next 25 years, electricity demands are expected to increase by 25% (EIA 2011), requiring more electricity generation and, generally, more water and land. Water availability is increasingly being considered in siting of new power plants and in licensing decisions, as are the impacts of associated land use for energy development on wildlife. Recognizing the importance of these issues in long-term siting of power plants, and also the transmission lines necessary to accommodate new electricity build outs, the Western Governors' Association (WGA) and Western Electric Coordinating Council, are working with teams from DOE and the National Laboratories to design scenarios for transmission planning that incorporate water availability and wildlife habitat. Additional power will be needed in the West to accommodate growing populations, and to support development of renewables in regions lacking transmission. The project is using different projected fuel portfolios to evaluate potential siting options for new power plants, in order to design optimal transmission line designs. In light of variability in water availability, the resilience of the system to drought is being incorporated into the planning (Iseman and Schroeder 2011). Further, funding from DOE is being used by the Western Governors' Association (Wildlife Council) to coordinate and create the Western Wildlife Crucial Habitat Assessment Tool (CHAT). One of the major goals of the CHAT, scheduled for release in 2013, is to "improve analysis of landscape-scale energy, land use and transportation projects as well as land conservation and climate adaptation strategies by providing prioritized information"(WGWC 2011). As of February 2012, five states have released their web-based mapping programs designed to display crucial wildlife habitat and corridors across the West:

Arizona, HabiMap™ Arizona
California, ACE II
Montana, CAPS
Southern Great Plains States, Southern Great Plains CHAT
Washington, PHS on the Web

2.2.2.4   Effects of Land on Energy Use

Land demands on energy resources include the need for energy, power, and combined heat and power (CHP) to support communities, towns, cities, municipalities, major urban centers, and associated industrial, commercial, and residential elements. Also included would be the need for production and delivery of electrical power and fuels to support conventional, electric, and hybrid transportation vehicles (ground and air) and mass transit. Also needed are energy, power, and CHP to support agriculture, aquaculture, and forestry operations and industry that includes biomass, biopower, and biofuels production.

Energy is also needed to support operations on land used for mines, wells, and related infrastructure conducting energy mineral extraction, processing, refining, transport, storage, and distribution.

Climate variation and change impacts would include extreme weather events that can damage infrastructure and disrupt operations. Droughts and elevated temperatures that reduce surface water supplies and watershed flow timing would also affect hydropower and thermopower generation and the production and processing of biomass for bioenergy. Energy in support of land use can also influence climate through the combustion of fuels and increased evaporation (from hydropower reservoir), evapotranspiration, and emission of other soil GHGs, particulates, and aerosols. Climate change influences energy through its effects on demand, distribution, patterns and intensity of use, and types of energy available (Dale et al. 2011). Examples include the increasing demands for air conditioning and heating as temperatures rise and fall. The types and energy use intensity of technologies used will also vary with climate, ranging from evaporative cooling and electrical heat pump systems for more moderate cooling and heating to refrigeration cycle based air conditioning and fossil fuel (gas, oil, coal) fired furnaces to meet more demanding cooling and heating needs. Finally, the supply chain for petroleum and other fossil fuels relies on coastal and major inland river ports and nearby refineries that are subject to disruption from increased storm intensity (hurricanes, tornados, floods, high seas, storm surge, and sea level rise) brought by climate change.

2.2.2.5   Effects of Water on Land Use

Water interacts with land through the hydrological cycle, when stored on land, and along coastal zones. Hydrological fluxes include precipitation, runoff, evapotranspiration, percolation (shallow, deep), and stream flow (Oki et al. 2006). Water can be stored on land surfaces in snowpacks, lakes, wetlands, and dams; in soil within the vadose zone; and deep below ground as groundwater. Humans have significantly changed the hydrological cycle by building dams, diverting surface water, and extracting groundwater.

While past development and protection of water resources in the U.S. has helped protect ecosystems while satisfying our needs, it is uncertain how future demands for freshwater will be met as the climate changes and the population grows (Vörösmarty et al. 2004; Rosegrant et al. 2009). There are at least four issues associated with water and its effect on land under climate change conditions.

The first issue is increased competition among water uses. In the U.S., about 80% of water withdrawals are for thermoelectric power (49%) and irrigation (31%). The rest is for public supply (11%), industrial (4%), aquaculture (2%), mining (1%), domestic (1%), and livestock (<1%) uses (Kenny et al. 2009). Consumptive water use is approximately 28-29% of total withdrawals on a national average, based on estimates for 1995 (Solley et al. 1998). Of this consumptive use, agriculture accounts for approximately 80-84% on a national average, with regional variations (Solley et al. 1998; Gollehon 2012). Regional changes in precipitation, depleted groundwater (e.g., Ogallala aquifer), and reduced snowpacks (e.g., Columbia Basin) are expected to affect future water availability and spur competition for water resources. Changes in snowpack will also influence the timing of water availability (IPCC 2008).

The second issue is the constraints that water might impose on future uses and management of land, primarily for food and biofuel production. Changes—especially reductions—in precipitation will require adapting cropping practices to new climatic conditions. This means more reliance on and more efficient use of "green" water (precipitation and transpiration) and less reliance on "blue" water (aquifers, lakes, and reservoirs) (Falkenmark and Rockström 2006). Farms will have to use water more efficiently by managing soil disturbance, plant populations, and nutrient pools (Hatfield et al. 2001; Turner 2004; Passioura and Angus 2010) and selecting cultivars that use water more efficiently (Morison et al. 2008). In developing biofuel landscapes, it will be necessary to examine the water footprint characteristics of biofuel crops (Bhardwaj et al. 2011) and the climate feedbacks of biofuel land covers (Loarie et al. 2011).

The third issue is the possible effects of altered hydrological regimes on soil biogeochemical cycles. Changes to soil moisture regimes, especially those brought on by longer or more pronounced dry or wet periods, will affect future soil emissions of carbon dioxide, nitrous oxide, and methane.

The final issue involves the effects of extreme events (e.g., droughts and floods) on agricultural systems, unmanaged systems, and developed areas. Two examples are the 2011–2012 Texas drought and the 1993 Upper Mississippi River Basin flood, which both brought significant economic, social, and environmental consequences (ref).

At the margins of the water-land interface are the coastal and island regions, which are vulnerable to major changes in sea state conditions. In these areas—which increasingly represent some of the most populated and infrastructure-dense areas in the U.S.—climate change affects both sea level rise and the increasing frequency and intensity of storms (Karl et al. 2009). Based on U.S. Census data (Census

2012), total national population in 2010 was nearly 309 million. Of this total, 48% live in the interior U.S. with an average national population density of 56 per square mile (sqm), while 52% live along coastlines with an average overall density of 180 per sqm. Of the total US population, nearly 23% are along the Atlantic Coast with an average population density of 475 per sqm; nearly 9% are along the Great Lakes with 236 per sqm; about 7% are along the Gulf of Mexico with 181 per sqm; and 13% are along the Pacific Coast, Alaska, and Hawaii with 81 per sqm. Within these coastal regions, population density ranges from a low of 2 per sqm for Alaska to a high of 9,800 per sqm for the District of Columbia. Despite being on the margins of the land-water interface, the coastal land-water-climate interactions, especially in the higher population density coastal marine waters, will have potentially strong climate change linkages that can have major impacts on land use and infrastructure in these regions.

2.2.2.6   Effects of Land on Water Use

The conterminous U.S. covers over 3 million square miles of land and water. Land use categories include cropland (23%), pasture and range (31%), forest (30%), urban (3%), and other areas (3%). The patterns of land use in upcoming decades, and thus the demands for water resources, will be determined largely by societal development and climate change. Land use and land use changes can affect water resources through, for example, interactions with groundwater (Scanlon et al. 2005), alterations in albedo and evapotranspiration (Loarie et al. 2011), and the productivity and water-use efficiency of future crops (Izaurralde et al. 2003). Meeting future water demands with dwindling water resources will require collaboration across disciplines and an understanding of the connections among the atmosphere, hydrosphere, lithosphere, biosphere, and anthroposphere (Wagener et al. 2010).

Current policies on energy security have opened a path for the development of lignocellulosic biofuels (EISA 2007; NAS 2011b). Large tracts of land will be required to meet lignocellulosic ethanol production targets (DOE 2011). The design of sustainable lignocellulosic biofuel landscapes will require a solid understanding of, among other things, their water footprint characteristics (i.e., consumptive water use per unit of energy produced) (Bhardwaj et al. 2011) and climate feedbacks (Loarie et al. 2011). Increasing food production to satisfy future national and international markets will depend, in part, on the ability to use crop water more efficiently by developing new cultivars and implementing water-efficient cropping designs (e.g., water-stress tolerant cultivars, subsurface irrigation).

Natural ecosystems provide many environmental services, such as purifying water, mitigating floods and droughts, and protecting coastal shores from wave erosion (Daly et al. 1997). While managed ecosystems provide vital goods for human well-being (e.g., food, fiber, shelter), they are less able to provide water-related ecosystem services (e.g., increased runoff, reduced water quality). Biofuels may offer an opportunity to rethink the organization of future landscapes based on sustainability and climate mitigation. From the standpoint of land effects on water, this would mean increased diversification of production landscapes to improve water-related ecosystem services. However, the impact of biofuels on water resources will also depend on the degree of scale-up, the regions of the country where it is done, and the technical approaches used (NAS 2007).

2.3   The Integrated Energy-Land-Water-Climate Nexus

An important goal for goal this report is to provide decision-makers with a clear understanding that the convergence of supply and demand issues related to energy, water, and land, within the context of a changing climate, will be central to making informed plans and policy choices for climate change mitigation and adaptation. This argues for taking an integrated, risk-based approach to climate assessment that considers the impacts of climate variability and change within the climate-EWL Nexus. This approach can also benefit by considering regional differences and multi-region interactions (Warren 2011; Dale - in press). Illustration 2.5 describes an effort underway to apply an integrated modeling to investigate climate-EWL interactions in the Southwestern U.S. This perspective is further explored and utilized through case study examples, scenarios, and strategies for mitigation and adaptation that will be introduced and addressed in greater detail later in this report.

Illustration 2.5: Water-Energy-Land-Climate Interactions with an Integrated Modeling System in the Southwestern U.S.

Given the potential complexity of climate-EWL interactions, it is important to identify the strongest and most policy-relevant relationships in order to develop tractable and useful models to assist decision making. A NOAA Societal Applications Research Program (SARP) project is building a modeling framework to capture key features of the climate-EWL interface at policy-relevant spatial scales. The models will help to quantify the tradeoffs and relationships among a comprehensive but streamlined set of water-energy constituents. This project takes advantage of two legacy resource modeling platforms—the Water Evaluation and Planning (WEAP) and the Long-range Energy Alternatives Planning (LEAP)—by linking them through data sharing of specific climate-EWL variables (Stockholm Environment Institute 2012). The initial area of interest is the Southwestern U.S., with a particular emphasis on California.

On the water-land-climate side of the ledger, WEAP starts with a monthly, climatically driven simulation of the hydrologic cycle, where multiple "catchments" are used to represent the spatial and temporal attributes of climate on the watershed. The WEAP model covers most of the Southwestern U.S. in order to capture the dependence of California on imported water from the Colorado River Basin. The region is broken into more than 300 catchment objects based on elevation and broad land-use classes, including native, urban, and irrigated agriculture types. This region is shown in Figure 2.9, with the inset graph illustrating some details of the model in Northern California.

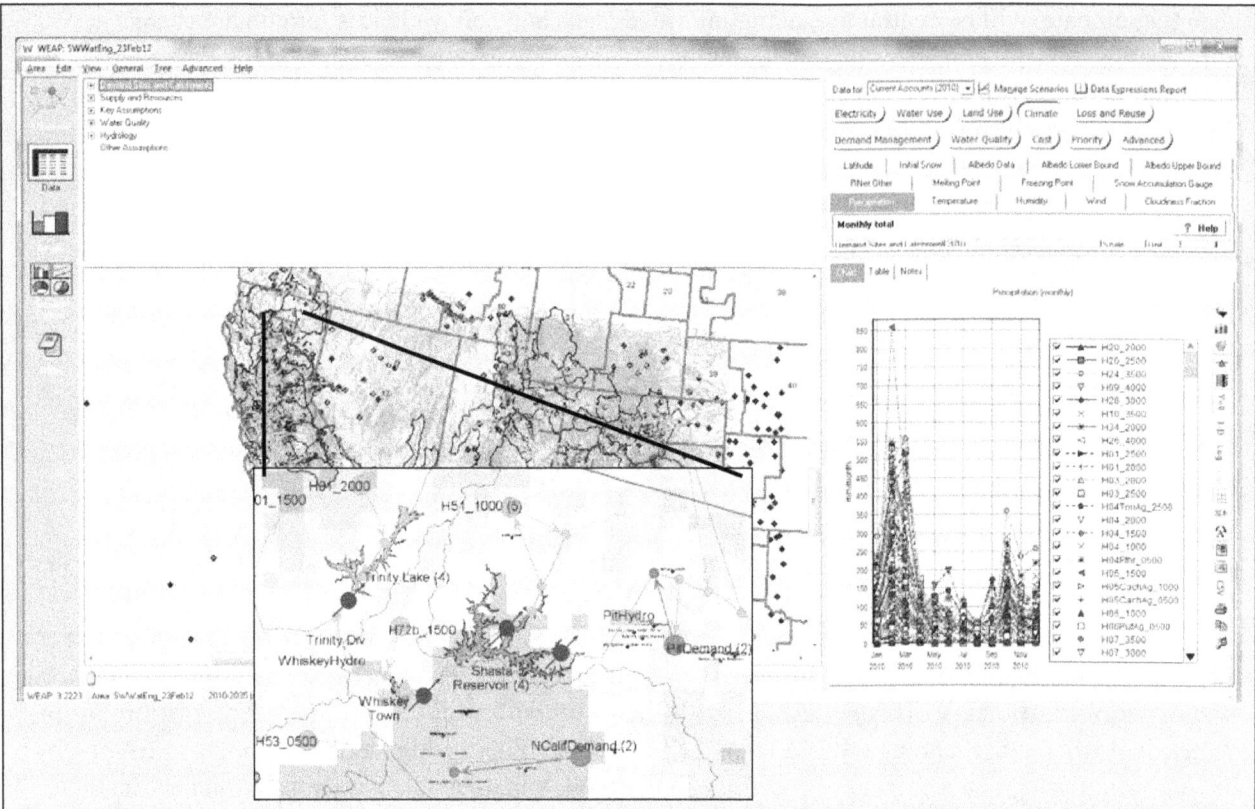

Figure 2.9. Screen shot of the WEAP model for the Southwestern U.S., with the inset graph showing a region in Northern California that includes Shasta, Trinity, and Whiskeytown Reservoirs (green triangles), a set of catchment objects (green dots), water demands (red dots), non-reservoir hydropower (blue rectangles), and observed stream gage locations (blue circles). Graph on the bottom-right is monthly precipitation for the catchment objects for the first year of simulation.

On the energy side of the ledger, this LEAP model focuses only on the electricity energy sector. Electricity demand is separated into non-water electricity as a per-capita demand multiplied by the population of California; and water-sector electricity, which includes water treatment and end-use, groundwater pumping, and water conveyance. Total water-related electricity demand is estimated directly in WEAP, as WEAP simulates municipal and agricultural water delivery and thus tracks the volume of treated water and implied end-use electricity demand; the volume of groundwater pumped; and the volume of water conveyed. Total water-related electricity demand is simply the sum of these delivery volumes times their specific electric intensity rates in kWh/m^3. Figure 2.10 shows the elements of the WEAP-LEAP integrations that are explicitly considered.

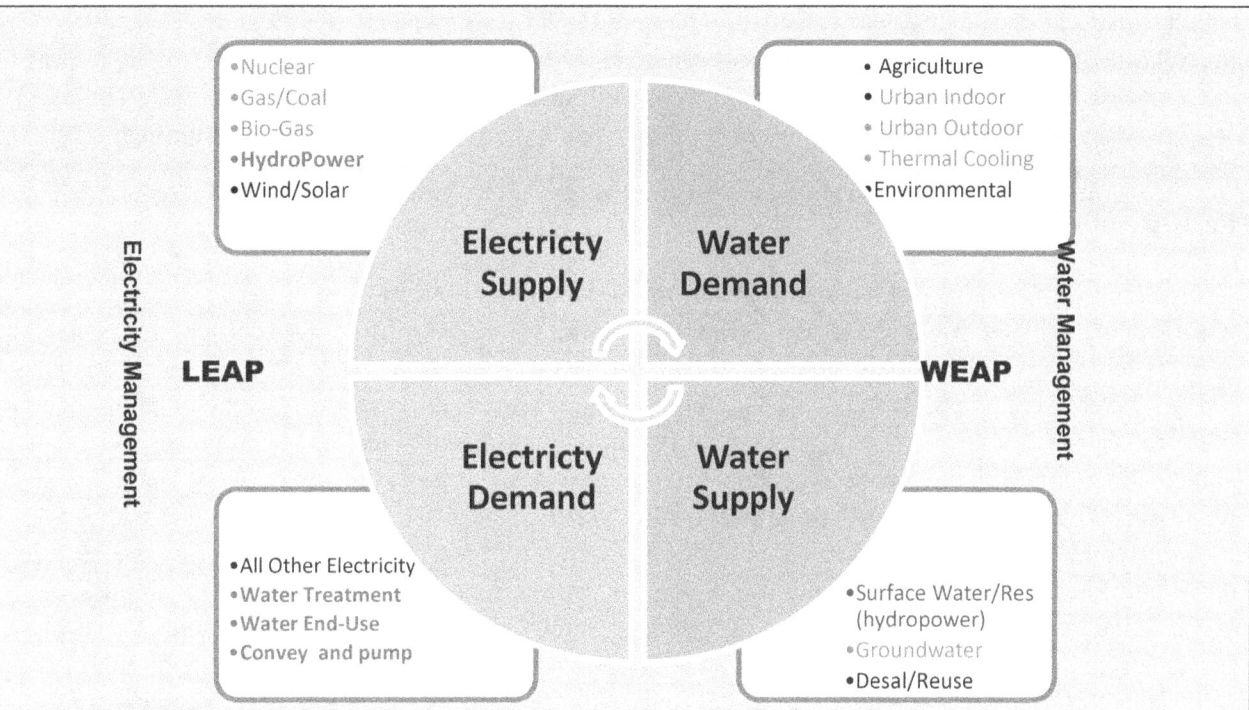

Figure 2.10. Water and electricity elements represented within the WEAP-LEAP integrated model. Colored text represent cross model dependencies. For example, while Hydropower generation (blue) is accounted for by LEAP in electricity dispatch, its availability is simulated in WEAP. Thermal cooling requirements are based on electricity demand from LEAP, but the amount of water available is estimated by WEAP.

WEAP catchments generate streamflow, which feeds the major river systems of the Southwestern U.S. These rivers extend from the South Platte River in Colorado, the Coastal Rivers of California, the Salt River of Arizona, the Klamath River of Southern Oregon, and 60 others. This river network includes the primary storage reservoirs and water transfer projects. The model represents 50 of the largest reservoirs and lakes, with a combined storage capacity of nearly 100 million acre-feet and more than 50,000 gWh of hydropower annually. WEAP estimated hydropower is passed to LEAP for dispatch. Water demands in WEAP are spatially explicit, and include the primary urban regions of the Southwest such as the Colorado Front Range, the Salt Lake Valley, the Phoenix Metropolitan Region, etc. In California, water demand is represented by more than a dozen of largest municipal regions, such as the Bay Area, the Central Valley, San Diego, the LA Basin, the Santa Clara Basin, etc. In the current version of the regional model, indoor water demands are computed on a per-capita basis, while both agricultural and urban outdoor water demands are calculated based on acreage estimates and soil moisture deficits. In addition to hydropower, WEAP estimates the electricity needed by water related activity, including conveyance, groundwater pumping, and municipal use (potable treatment, end-use requirements such as water heating, and wastewater treatment).

There are five competing water demands represented in the model, each given an integer priority according to the WEAP allocation logic, with the highest priority assigned a value of 1. From highest (1) to lowest (5) priority, these are environmental flows (such as the Bay-Delta), thermoelectric cooling, urban indoor, urban outdoor, and irrigated agriculture. Under water short conditions, WEAP first allocates to the highest priority, then the second highest priority, and so on. The model was configured to simulate over the period 2010 through 2035, with an assumed 1.8% population growth rate for the Southwest region. For simplicity, we have assumed that the climate of 1975 through 2000 simply repeats itself for this future period. This period of record includes the two strong drought episodes of 1976 to 1977 (corresponding to 2011 and 2012) and the prolonged 5-year drought spanning 1988 through 1992 that correspond to years 2023 through 2027, for this future projection out to 2035.

Early Results

   Figure 2.11 shows the results for a Business-As-Usual (BAU) projection out to 2035, showing electric demand by the water sector (top-left), total electricity generation by source (bottom-left), water delivery to the agriculture and municipal sectors (top-right), and supply delivery by water source (bottom-right). Both bottom graphs include the CO_2 emission equivalent for all electricity generation and suggests greater emissions from electricity generation during drought attributed to reduced hydropower generation and greater groundwater pumping. The intensive drought leads to substantial reductions in water delivery to the agricultural sector, primarily from local surface water sources. The results suggest that without new capacity California will have to import more electricity from outside the state, leading to greater CO_2 emissions, and under the strong drought conditions there is greater reliance on groundwater supplies, reduced delivery to the agriculture sector, and greater overall CO_2 emissions. While in some regions of the country, the importance of freshwater availability for cooling thermometric plants is important, in California, this represents less than 1% of total water withdrawals, as cooling water requirements are largely met by ocean water.

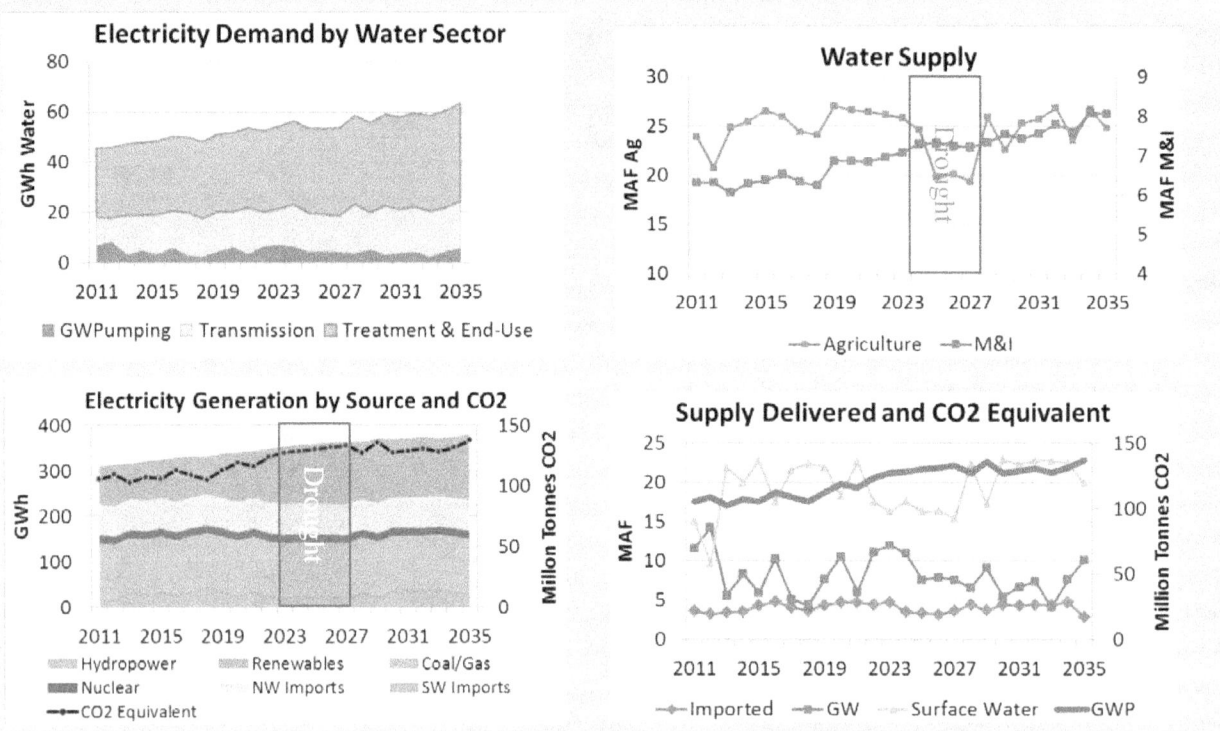

Figure 2.11. Electricity demand for each water-related activity including groundwater pumping, treatment and end-use, and transmission (top-left). Electricity generation by source and their CO_2 equivalent (bottom –left). Water supply delivered to the agriculture and municipal and industrial sectors (top-right), and water supply delivered by source and the equivalent CO_2 electricity equivalent for that level of generation.

2.4 References

Atlantic Council. 2011. Energy for water and water for energy, A report on the Atlantic Council's workshop: How the Nexus Impacts Electric Power Production in the United States", October 2011, ISBN: 978-1-61977-001-0.

Averyt K, J Fisher, A Huber-Lee, A Lewis, J Macknick, N Madden, J Rogers, and S Tellinghuisen. 2011. Freshwater use by U.S. power plants: Electricity's thirst for a precious resource, A report of the energy and Water in a Warming World initiative. Union of Concerned Scientists, Cambridge, MA. November.

Bales J, N Gollehon, and C Bernacchi. 2011. "Sustainable Feedstocks for Advanced Biofuels: Water Resource Impacts of Feedstock Production and Conversion," Chapter 4 of *Sustainable Alternative Fuel Feedstock Opportunities, Challenges, and Roadmaps for Six U.S. Regions*, Karlen, D. and Steiner, J (Eds), Soil and Water Conservation Society.

Berndes G. 2002. "Bioenergy and water – the implications of large-scale bioenergy production for water use and supply." *Global Environ. Change* 12: 253-271.

Beshears E. 2007. "Obstacle to more power: Hot water: River temperature so high that Duke Energy curtails work at 2 plants," *Charlotte Observer*, August 12.

Bhardwaj AK et al. 2011. "Water and energy footprints of bioenergy crop production on marginal lands." *Global Change Biology – Bioenergy* 3(3):208-222.

BOR. 2011. SECURE Water Act Section 9503© - Reclamation Climate Change and Water 2011, Report to Congress by the U.S. Department of Interior Bureau of Reclamation, April 2011.

Brekke LD et al. 2009. Climate change and water resources management—A federal perspective, U.S. Geological Survey Circular 1331, 65 pp. Accessed at: http://pubs.usgs.gov/circ/1331/

Carter NT. 2011. "Energy's water demand: Trends, vulnerabilities, and management," Congressional Research Service Report for Congress, R41507.

CBES. 2009. Land-use change and bioenergy: Report from the 2009 workshop, ORNL/CBES-001, U.S. DOE/EERE and ORNL, Center for Bioenergy Sustainability. Accessed at: http://www.ornl.gov/sci/besd/cbes.shtml.

CCSP. 2008. Effects of Climate Change on Energy Production and Use in the United States, Synthesis and Assessment Product 4.5, U.S. Climate Change Science Program, February 2008.

Census. 2012. The 2012 Statistical Abstract: Population in Coastal Counties – 1980 to 2010, The National Data Book. U.S. Census Bureau. Accessed at: http://www.census.gov/compendia/statab/cats/population.html

Colorado Water Conservation Board. 2010. State Wide Water Supply Initiative-2010. Accessed at: http://cwcb.state.co.us/water-management/water-supply-planning/Pages/SWSI2010.aspx.

Cook PS and J O'Laughlin. 2011. Forest Biomass Supply Analysis for Western States by County: Final Report to the Western Governors' Association. College of Natural Resources, University of Idaho, January 2011.

Cooley H et al. 2011. Water for Energy: Future Water Needs for Electricity in the Intermountain West, Pacific Institute, Oakland, CA.

Crutzen PJ and EF Stroemer. 2000. "The Anthropocene." *International Geosphere-Biosphere Programme Newsletter* 41:17-18.

Daily GC, S Alexander, PR Ehrlich, L Goulder, J Lubchenco, PA Matson, HA Mooney, S Postel, SH Schneider, D Tilman, and GM Woodwell. 1997. "Ecosystem services: Benefits supplied to human societies by natural ecosystems." *Issues in Ecology* 2, Spring 1997.

Dale VH et al. (In press: draft chapter dated Oct 2011). "Modeling for integrating science and management," A chapter in *Land Use and the Carbon Cycle: Science and Applications in Coupled Natural-Human Systems* (eds. D Brown and D Robinson), Cambridge University Press.

Dale VH et al. 2011. "The land use – climate change – energy nexus." *Landscape Ecology* 26:755-773.

DOE. 2006. Energy Demands on Water Resources, Report to Congress on the Interdependency of Energy and Water, U.S. Department of Energy.

DOE. 2011. "U.S. Billion Ton Update: Biomass Supply for a Bioenergy and Bioproducts Industry, RD Perlack and BJ Stokes (Leads). ORNL/TM-2011/224, Oak Ridge National Laboratory, Oak Ridge, TN, August 2011, 227 p.

Dominguez-Faus R et al. 2009. "The Water Footprint of Biofuels: A Drink or Drive Issue?" *Environmental Science and Technology* 43:3005-3010.

Donner SD et al. 2008. "Corn-based ethanol production compromises goal of reducing nitrogen export by the Mississippi River." *PNAS* 105(11):4513-4518.

DSB. 2011. Trends and Implications of Climate Change for National and International Security, Task Force Report - Defense Science Board, Office of Undersecretary of Defense for Acquisition, Technology, and Logistics, Washington, DC, October 2011.

Ebinger J et al. 2011. Climate impacts on energy systems: Key issues for energy sector adaptation, Energy Sector Management Assistance Program, The World Bank, Washington, DC.

EIA - Energy Information Administration. 2011. Annual Energy Outlook 2011 with Projections to 2035. Report DOE/EIA-0383(2011), Department Of Energy, Release Date: December 16, 2010. Accessed at: http://www.eia.gov/oiaf/aeo/gas.html.

EISA. 2007. Energy independence and security act of 2007. H.R.6. In: 110th Congress, public law no: 110–140; 2007 [December 19].

Ellis EC, KK Goldewijk, S Siebert, D Lightman, and N Ramanjutty. 2010. "Anthropogenic transformation of the biomes, 1700 to 2000." *Global Ecology and Biogeography* 19:589-606.

EPA. 2011. "Water quality standards for surface waters." U.S. Environmental Protection Agency. Accessed at: http://water.epa.gov/scitech/swguidance/standards

EPRI. 2011. Water Use for Electricity Generation and Other Sectors: Recent Changes (1985-2005) and Future Projections (2005-2030). Technical Report 1023676, Electric Power Research Institute, November 2011.

Falkenmark M et al. 2005. "Consumptive water use to feed humanity – curing a blind spot." *Hydrology and Earth Systems Sciences* 9:15-28.

Falkenmark M. 2003. "Freshwater as shared between society and ecosystems: from divided approaches to integrated challenges," *Phil. Trans. R. Soc. Lond. B* 358:2037-2049.

Falkenmark M and J Rockström. 2006. "The new blue and green water paradigm: Breaking new ground for water resources planning and management." *J. Water Resour. Plann. Manage.* 132:129–132.

Feddema JJ, KW Oleson, GB Bonan, LO Mearns, LE Buja, GA Meehl, and WM Washington. 2005. "The importance of land-cover change in simulating future climates," *Science* 310: 1674-1678.

Ferguson IM et al. 2011. "Hydrologic and Land-Energy feedbacks of agricultural water management practices." *Environ. Res. Lett.* 6 (7pp.) 014006.

FIPCCWDI. 2011. Report to Congress—Strengthening the scientific understanding of climate change impacts on freshwater resources of the United States. Federal Interagency Panel on Climate Change and Water Data and Information, 49 p.

GAO. 2009. Energy-Water Nexus: Many Uncertainties Remain about National and Regional Effects of Increased Biofuel Production on Water Resources. U.S. Government Accountability Office Report to the Chairman, Committee on Science and Technology, House of Representatives, GAO-10-116, November 2009.

GAO. 2011. Energy-Water Nexus: Amount of Energy Needed to Supply, Use, and Treat Water is Location-Specific and can be Reduced by Certain Technologies and Approaches, U.S. Government Accountability Office Report to the Ranking Member, Committee on Science, Space, and Technology, House of Representatives, GAO-11-225, March 2011.

GAO. 2012. Energy-Water Nexus: Information on the Quantity, Quality, and Management of Water Produced during Oil and Gas Production, U.S. Government Accountability Office Report to the Ranking Member, Committee on Science, Space, and Technology, House of Representatives, GAO-12-156, January 2012.

Gerbens-Leenes W, AY Hoekstra, and TH van der Meer. 2009. "The Water Footprint of Bioenergy", *PNAS* 106 (25):10219-10223, June 23, 2009.

GLC. 2011. Integrating energy and water resources decision making in the Great Lakes Basin: An examination of future power generation scenarios and water resources impacts, A report of the Great Lakes Energy-Water Nexus Team, the Great Lakes Commission, October 2011.

Gollehon N. 2012. Agricultural Water Usage: Trends, Indicators, and What it all Means, USDA/NRCS, featured presentation for the Horinko Group 2012 Water Division Webinar 16-Feb-2012, Accessed at: http://www.thehorinkogroup.org/pubs/Feb162012WebinarPDF.pdf.

Gordon L et al. 2003. "Land cover change and water vapour flows: learning from Australia," *Phil. Trans. R. Soc. Lond. B* 358:1973-1984.

Groom MJ et al. 2008. "Biofuels and Biodiversity: Principles for creating better policies for biofuel production." *Conservation Biology* 22(3):602-609.

Gurdak et al. 2007. "Climate controls on unsaturated water and chemical movement in High Plains aquifer." *Vadose Zone Journal* 6:533-547.

Hatfield JL, TJ Sauer, and JH Prueger. 2001. "Managing soils to achieve greater water use efficiency: A review." *Agron. J.* 93:271–280.

Hester ET and MW Doyle. 2011. "Human impacts to river temperature and their effects on biological processes: A quantitative synthesis." *Journal of the American Water Resources Association* 47(3):571–587.

Hightower M et al. 2008. "The energy challenge." *Nature* 452:285-286.

IPCC. 2007. Climate Change 2007: Synthesis Report. Contribution of Working Groups I, II and III to the Fourth Assessment Report of the Intergovernmental Panel on Climate Change [Core Writing Team, Pachauri, R.K and Reisinger, A. (eds.)]. IPCC, Geneva, Switzerland, 104 pp.

Iseman T and A Schroeder. 2011. "Integrated planning: transmission, generation and water in the Western states," in *The Water-Energy Nexus in the American West* (D.S. Kenney and R. Wilkinson eds.), Edward Elgar Publishing, Inc. Cheltenham, UK.

Izaurralde, RC, NJ Rosenberg, RA Brown, and AM Thomson. 2003. "Integrated assessment of Hadley Centre climate change projections on water resources and agricultural productivity in the conterminous United States. II. Regional agricultural productivity in 2030 and 2095." *Agric. For. Meteor* 117:97-122.

Karl TR, JM Melillo, and TC Peterson (eds.). 2009. Global Climate Change Impacts on the United States. Cambridge University Press.

Kenny JF, NL Barber, SS Hutson, KS Linsey, JK Lovelace, and MA Maupin. 2009. Estimated Use of Water in the United States in 2005, U.S. Geological Survey Circular 1344. Reston, VA.

King CW. et al. 2008. "Water intensity of transportation." *Environmental Science & Technology* 42(21):7866-7872.

Kobos PH et al. 2011. "Combining power plant water needs and carbon dioxide storage using saline formations: Implications for carbon dioxide and water management practices." *Int'l Journal of Greenhouse Gas Control* 5:899-910.

Kocoloski, M et al. (2009). "Indirect land use change and biofuel policy." *Environ. Res. Lett.* 4 (5pp), 034008, downloaded on 05/04/2010.

Koshel P and K Mcallister (Rapporteurs). 2008. Expanding Biofuel Production: Sustainability and the Transition to Advanced Biofuels: Summary of a Workshop. The National Academies. Accessed at: http://www.nap.edu/catalog/12195.html.

Koshel P and K Mcallister (Rapporteurs). 2010. Transitioning to Sustainability Through Research and Development on Ecosystem Services and Biofuels: Workshop Summary. The National Academies. Accessed at: http://www.nap.edu/catalog/12806.html.

Loarie SR, DB Lobell, GP Asner, Q Mu, and CB Field. 2011. "Direct impacts on local climate of sugarcane expansion in Brazil." *Nature Climate Change* 1:105-109.

Lovich JE and JR Ennan. 2011. "Wildlife Conservation and Solar Energy Development in the Desert Southwest, United States." *BioScience* 61(12):982-992.

MacKnick J, R Newmark, G Heath, and KC Hallet. 2011. A review of operational water consumption and withdrawal factors for electricity generating technologies. National Renewable Energy Laboratory, Golden, Colorado. Accessed at: http://www.nrel.gov/docs/fy11osti/50900.pdf.

Madden NT. 2010. In hot water: Thermoelectric power and thermal pollution. Poster session presented at Energy-Water Interdependence, meeting of the American Geophysical Union, December 13–17, San Francisco, California.

Mahowald N. 2011. "Aerosol Indirect Effect on Biogeochemical Cycles and Climate." *Science* 334:794-796.

Marland G, RA Pielke, M Apps, R Avissar, RA Betts, KJ Davis, PC Frumhoff, ST Jackson, LA Joyce, P Kauppi, J Katzenberger, KG MacDicken, RP Neilson, JO Niles, DS Niyogi, RJ Norby, N Pena, N Sampson, and Y Xue. 2003. "The climatic impacts of land surface change and carbon management, and the implications for climate-change mitigation policy," *Climate Policy* 3(2):149-157.

McDonald RI, J Fargione, J Kiesecker, WM Miller, and J Powell. 2009. "Energy Sprawl or Energy Efficiency: Climate Policy Impacts on Natural Habitat for the United States of America." *PLoS ONE* 4(8):e6802, DOI:10.1371/journal.pone.0006802.

McLaughlin SB and ME Walsh. 1998. "Evaluating environmental consequences of producing herbaceous crops for bioenergy." *Biomass and Bioenergy*, 14(4):317-324.

McMahon JE and SK Price. 2011. "Water and Energy Interactions," *Annual Reviews of Environment and Resources* 36:163-191, DOI: 10.1146/annurev-environ-061110-103827.

MEA. 2005. *Ecosystems and Human Well-being: Synthesis.* Millennium Ecosystem Assessment, Island Press, Washington, DC.

Michael J, T Thompson, and J Sieber. 2012. "Climate Predictions: The Influence of Nonlinearity and Randomness." *Philosophical Transactions of the Royal Society A*, 370:1007-1111.

Miller SA 2010. "Minimizing Land Use and Nitrogen Intensity of Bioenergy." *Environmental Science & Technology* 44(10):3932-3939, DOI: 10.1021/es902405a.

Morison, JIL, NR Baker, PM Mullineaux, and WJ Davies. 2008. "Improving Water Use in Crop Production." *Philosophical Transactions of the Royal Society B-Biological Transactions* 363(1491):639–658, DOI: 10.1098/rstb.2007.2175.

NAS – National Academy of Sciences. 2007. Water implications of biofuels production in the United States. Washington, DC.

NAS – National Academy of Sciences. 2010a. Adapting to the Impacts of Climate Change. Washington, DC.

NAS – National Academy of Sciences. 2011a. America's Climate Choices. Washington, DC.

NAS – National Academy of Sciences. 2011b. Renewable Fuel Standard: Potential Economic and Environmental Effects of the U.S. Biofuel Policy. National Research Council of the National Academies, National Academies Press, Washington, DC.

NAS – National Academy of Sciences. 2010b. Advancing the Science of Climate Change. Washington, DC.

NASA – National Aeronautics and Space Administration. 2011. Ancient Dry Spells Offer Clues About the Future of Drought. Accessed December 2011 at: http://www.nasa.gov/topics/earth/features/ancient-dry.html

NatureServe. 2011. Ecological Systems of the United States. Accessed at: http://www.natureserve.org/index.jsp.

NETL – National Energy Technology Laboratory. 2010. 2010 Carbon Sequestration Atlas of the United States and Canada. 3rd ed. National Energy Technology Laboratory. Accessed at: http://www.netl.doe.gov/technologies/carbon_seq/refshelf/atlasIII/2010atlasIII.pdf

Newell CJ and JA Connor. 2006. "Strategies for Addressing Salt Impacts of Produced Water Releases to Plant, Soil, and Groundwater." Energy API Publication 4758, Sept 2006.

NRC – Nuclear Regulatory Commission. 2007. Power Reactor Status Reports, July 2007. Nuclear Regulatory Commission. Accessed at: http://www.nrc.gov/reading-rm/doc-collections/event-status/reactor-status/

NRC – Nuclear Regulatory Commission. 2010. Power Reactor Status Reports, July–August 2010. Nuclear Regulatory Commission. Accessed at: http://www.nrc.gov/reading-rm/doc-collections/event-status/reactor-status

NRC – Nuclear Regulatory Commission. 2011. Power Reactor Status Reports, July 2011. Nuclear Regulatory Commission Accessed at: http://www.nrc.gov/reading-rm/doc-collections/event-status/reactor-status

Oki T and S Kanae. 2006. "Global Hydrological Cycles and World Water Resources." *Science* 313(5790):1068-1072, DOI: 10.1126/science.1128845.

Outka U. In Press. "The Energy-Land Use Nexus: Symposium Essay." *Journal of Land Use and Environmental Law*.

Outka U. 2011. "The Energy-Land Use Nexus: Symposium Essay." University of Kansas School of Law Working Paper No. 2011-08, Forthcoming in the Journal of Land Use and Environmental Law, 2012.

Passioura JB and JF Angus. 2010. "Improving Productivity of Crops in Water Limited Environments." *Advances in Agronomy* 106:37–75, DOI: 10.1016/S0065-2113(10)06002-5.

Pate R, M Hightower, C Cameron, and W Einfeld. 2007. Overview of Energy-Water Interdependencies and the Emerging Energy Demands on Water Resources. SAND 2007-1349C, Sandia National Laboratories, Albuquerque, New Mexico.

Pielke RA 2005. "Land Use and Climate Change." *Science* 310(5754):1625-1626, DOI: 10.1126/science.1120529

Reilly J and S Paltsev. 2007. Biomass Energy and Competition for Land. MIT Report No. 145, Massachusetts Institute of Technology Joint Program on the Science and Policy of Global Change, Cambridge, Massachusetts.

Reilly TE, KF Dennehy, WM Alley, and WL Cunningham. 2008. Ground-Water Availability in the United States, U.S. Geological Survey Circular 1323, U.S. Geological Survey, Reston, Virginia

Rosegrant MW, C Ringler, and T Zhu. 2009. "Water for Agriculture: Maintaining Food Security under Growing Scarcity." *Annual Review of Environment and Resources* 34:205–222, DOI: 10.1146/annurev.environ.030308.090351

Scanlon BR, RC Reedy, DA Stonestrom, DE Prudic, and KF Dennehy. 2005. "Impact of Land Use and Land Cover Change on Groundwater Recharge and Quality in the Southwestern US." *Global Change Biology* 11(10):1577-1593, DOI: 10.1111/j.1365-2486.2005.01026.x

Scott C, M Pasqualetti, J Hoover, G Garfin, R Varady, and S Guhathakurta. 2009. Water and Energy Sustainability with Rapid Growth and Climate Change in the Arizona-Sonora Border Region: A Report to the Arizona Water Institute.

Scott CA and MJ Pasqualetti. 2010. "Energy and Water Resources Scarcity: Critical Infrastructure for Growth and Economic Development in Arizona and Sonora," *Natural Resources Journal* 50(3):645-682.

Scott CA, SA Pierce, MJ Pasqualetti, AL Jones, BE Montz, JH Hoover. 2011. "Policy and Institutional Dimensions of the Water-Energy Nexus," *Energy Policy* 39(10):6622-6630, DOI: 10.1016/j.enpol.2011.08.013.

Smith P, PJ Gregory, D van Vuuren, M Obersteiner, P Havlik, M Rounsevell, J Woods, E Stehfest, and J Bellarby. 2010. "Competition for Land." *Philosophical Transactions of the Royal Society B-Biological Sciences* 365(1554):2941-2957, DOI: 10.1098/rstb.2010.0127.

Smith, J. 2011. Brown's Ferry Nuclear Power Plant has to Shut Down Again. The Energy Collective, August 9, 2011. Accessed at http://theenergycollective.com/jcwinnie/62883/brown-s-ferry-nuclear-power-plant-has-shut-down-again.

Solley WB, RR Pierce, and HA Perlman. 1998. Estimated use of water in the United States in 1995, U.S. Geological Survey Circular 1200, 71 p.

Southwest Climate Change Network. 2012. Renewable Energy Supply and Transmission. Accessed at: http://www.southwestclimatechange.org/solutions/reducing-emissions/energy-supply-transmission

Stillwell AS, CW King, ME Webber, IJ Duncan, A Hardberger. 2009. Energy-Water Nexus in Texas. Texas State Energy Conservation Office; University of Texas at Austin; and Environmental Defense Fund.

Stockholm Environment Institute. 2012. Software. Accessed at: http://www.sei-us.org/weap.

Stone B. 2009. "Land Use as Climate Change Mitigation." *Environmental Science & Technology* 43(24):9052-9056, DOI: 10.1021/es902150g.

Texas Water Development Board. 2012. 2012 State Water Plan. Accessed at: http://www.twdb.state.tx.us/wrpi/swp/swp.asp.

Trask M. 2005. California's Water-Energy Relationship, Final Staff Report CEC-700-2005-011-SF Prepared in support of the 2005 Integrated Energy Policy Report Proceeding (O4-IEPR-01E), California Energy Commission, Nov 2005.

Turner NC. 2004. "Agronomic Options for Improving Rainfall-Use Efficiency of Crops in Dryland Farming Systems." *Journal of Experimental Botany* 55(407):2413–2425, DOI:10.1093/jxb/erh154.

Tyner WE, Taherpour F, Zhuang Q, Dirur D, and U Baldos. 2010. Land Use Changes and Consequent CO2 Emissions Due to U.S. Corn Ethanol Production: A Comprehensive Analysis, Final report. Department of Agricultural Economics, Purdue University, Richmond, Indiana.

U.S. Bureau of Reclamation. 2008. Southern Delivery System Final Environmental Impact Statement. December 2008.

Vine E. 2011. Adaptation of California's Electricity Sector to Climate Change." *Climatic Change* 111(1): 75-99, DOI: 10.1007/s10584-011-0242-2.

Voinov A and H Cardwell. 2009. "The Energy-Water Nexus: Why Should We Care?" *Journal of Contemporary Water Research & Education* 143(1):17-29, DOI: 10.1111/j.1936-704X.2009.00061.x.

Vörösmarty C, DP Lettenmaier, C Leveque, M Meybeck, C Pahl-Wostl, J Alcamo, W Cosgrove, H Grassl, H Hoff, P Kabat, F Lansigan, R Lawford, and R Naiman. 2004. "Humans Transforming the Global Water System." *EOS, Transactions, American Geophysical Union* 85(48):509–520, DOI: 10.1029/2004EO480001

Wagener T, M Sivapalan, PA Troch, BL McGlynn, CJ Harman, HV Gupta, P Kumar, PSC Rao, NB Basu, and JS Wilson. 2010. "The Future of Hydrology: An Evolving Science for a Changing World." *Water Resources Research* 46:W05301, DOI: 10.1029/2009WR008906.

Warren R. 2011. "The Role of Interactions in a World Implementing Adaptation and Mitigation Solutions to Climate Change." *Philosophical Transactions of the Royal Society A-Mathematical Physical and Engineering Sciences*, 369(1934):217-241, DOI: 10.1098/rsta.2010.0271.

WGA - Western Governors' Association. 2009. Western Renewable Energy Zones-Phase 1 Report. June 2009, pp. 23. Accessed at: http://www.westgov.org/rtep/219-western-renewable-energy-zones

WGA - Western Resource Advocates. 2008. Personal communication between Stacy Tellinghuisen, Western Resource Advocates, and Bob Peters, Denver Water. July 28, 2008.

WGWC - Western Governors' Wildlife Council. 2011. White Paper-Version II, Western Wildlife Crucial habitat Assessment Tool (CHAT):Vision, Definitions and Guidance for State Systems and Regional Viewer, August 2011. Accessed at: http://www.westgov.org/initiatives/wildlife.

Williams et al. 2011. "Alternatives for Decarbonizing Existing USA Coal Power Plant Sites." *Energy Procedia* 4:1843-1850.

Wise M, K Calvin, A Thomson, L Clarke, B Bond-Lamberty, R Sands, SJ Smith, A Janetos, and J Edmonds. 2009. "Implications of Limiting CO_2 Concentrations for Land Use." *Science* 324(5931):1183-1186, DOI: 10.1126/science.1168475.

Yi-Wen C, B Walseth, and S Suh, 2009. "Water Embodied in Bioethanol in the United States." *Environmental Science & Technology* 43(8):2688-2692, DOI: 10.1021/es8031067.

Zalasiewicz J, M Williams, A Haywood, and M Ellis. 2011. "The Anthropocene: A New epoch of Geological Time?" *Philosophical Transactions of the Royal Society A-Mathematical, Physical & Engineering Sciences* 369(1938):835-841, DOI: 10.1098/rsta.2010.0339.

Zalasiewicz J, M Williams, W Steffen, and P Crutzen, 2010. "The New World of the Anthropocene." *Environmental Science & Technology* 44(7):2228-2231, DOI: 10.1021/es903118j.

3.0 Observed and Projected Climate Impacts on Interfaces

The many bilateral interfaces explored in section 2 form a dynamic set of interacting processes related through a complex network of feedbacks in which the response can be immediate or lagged in time. As a simple example, consider the case of intensifying drought over long time periods. This change could initiate a cascade of natural and human-mediated responses across numerous bilateral interfaces. Natural responses might include vegetation change accompanied by reduced basin production of water (land-water) that in turn reduces reservoir volumes and hydropower production (energy-water). Efforts to adapt to more frequent and intense drought might involve replacing thermoelectric production with low-water-use renewables (energy-water) or reducing grazing densities to better manage fragile grasslands (land-water).

In this simple example it is important to note that any given sectoral or bilateral interface's response will both depend on and influence the responses of other bilateral interfaces. Reduced basin production due to changes in vegetation and grazing requires new reservoir management strategies that balance competing needs for hydropower, thermoelectric power, agriculture, municipal water uses, irrigation, recreation, and other environmental purposes. That is, allocating more water to one use means less water for other uses. This creates competition across the bilateral interfaces. The Technical Input Report on Climate Change and Infrastructure, Urban Systems and Vulnerabilities also recognizes such cascading impacts among interdependent infrastructures (NCA 2013a).

While section 2 provides a comprehensive view of the bilateral interfaces and their relation to climate, focusing on individual interfaces alone does not adequately capture the complexity and importance of the energy-water-land (EWL) system. This section explores how individual bilateral interfaces network and interact in response to specific climate change forcings. Just as describing each individual block that comprises a building will not convey the structure and form of the building, so to must we consider how the individual bilateral interfacial building blocks "assemble" in response to different climate forcings and the nuances of different physical and human settings.

3.1 Introduction of Example Case Studies and Illustrations

Section 2 introduced the many bilateral interfaces, and their even greater number of interrelationships. As such, it would be impossible to explore the full breadth of the EWL interfacial landscape. Instead, this section investigates several examples, constructed as discrete case studies according to two broad themes: (1) extreme climate events and (2) regional differences.

These themes were selected because they address key issues facing resource managers and policy makers. These themes also help demonstrate a few of the unique ways that the individual bilateral interfaces network and interact across the broad spectrum of the EWL system in response to different climate forcings. A unique feature of extreme climate events is the short time frame over which "extreme" changes are generally experienced. The second theme was selected to highlight regional differences. In particular, cross-sectoral response to climate are moderated by several interacting factors including differences in physiography of the region, built infrastructure, energy technologies, resource management strategies, and adaptation options to name a few.

The two themes are explored through one or more current cases studies. Each case study is introduced with a brief overview of the issue in terms of climatic forcing. Using the individual bilateral interfaces as the building blocks, the unique networking, interactions, and dynamics expressed across the interfaces is explored and discussed. For each case study, observed interfacial changes (current observations) are explored along with a review of potential changes forced by projected climate change.

Two "illustrations," or representative examples of the consequences of decision making on important aspects of EWL interactions, are also explored. Topics for the illustrations are (1) EWL implications of switching from coal to natural gas, and (2) decision-making considerations associated with carbon capture and storage (CCS).

3.2 Theme 1: Extreme Climate Events

Extreme events are often described as flood, drought, heat waves, tornadoes, hurricanes and high wind storms. Such events often represent the "worst-case scenario" in terms of resource management, infrastructure protection and regional planning processes. As such, past events are often used to assess resilience of a system. This is a common feature in water management where a drought of record (based on historical observations) is used to optimize water system operations. However, there is growing evidence that climate change will intensify the frequency, duration, and severity of extreme weather events (IPCC 2007; Karl et al. 2009). This raises questions as whether there is sufficient adaptive capacity to handle such change (Milly et al. 2008).

The expression of these events as well as the interaction across EWL interfaces vary by region. As such, it is beyond the scope of this report to fully characterize all extreme events for all regions. Rather, we focus on a single event, the co-occurring drought and heat wave that struck the Southern High Plains (with specific focus on Texas) in 2011. Whether this event is representative of "climate change" or natural climate variability is immaterial to this analysis. Rather, this event provides a concrete example of an extreme drought and heat wave, the worst on record for Texas in many respects. While extreme by today's standard, the severity of drought and the occurrence of heat waves across the United States are projected to occur more often over the next several decades. Thus, this case study provides insight for future co-occurring drought and heat waves and their linkages to energy, water and land resources. Additionally, observation and analysis of system interactions will provide indications about the resilience, adaptive capacity, and potential risks.

This section begins by introducing the 2011 Texas drought and heat wave. Concrete examples characterizing the impact of the drought and heat wave on the energy, water, and land interfaces are then explored, particularly considering how the impact at one interface will reinforce or moderate the affects at another. Along the way, various opportunities for future mitigation and adaptation are considered. To assist in visualizing these interfacial interactions in the context of our "resource demand, supply endowment, and technologies" conceptual model (Figure 2.3), key linkages and associated adaptation/mitigation options associated with the 2011 Texas drought and heat wave are presented in Figure 3.1. Finally, this section concludes with a brief discussion of how insights gained from Texas might be extended to other regions. Please note that extreme events are also briefly addressed in the discussion of the regional differences theme in section 3.3.4.

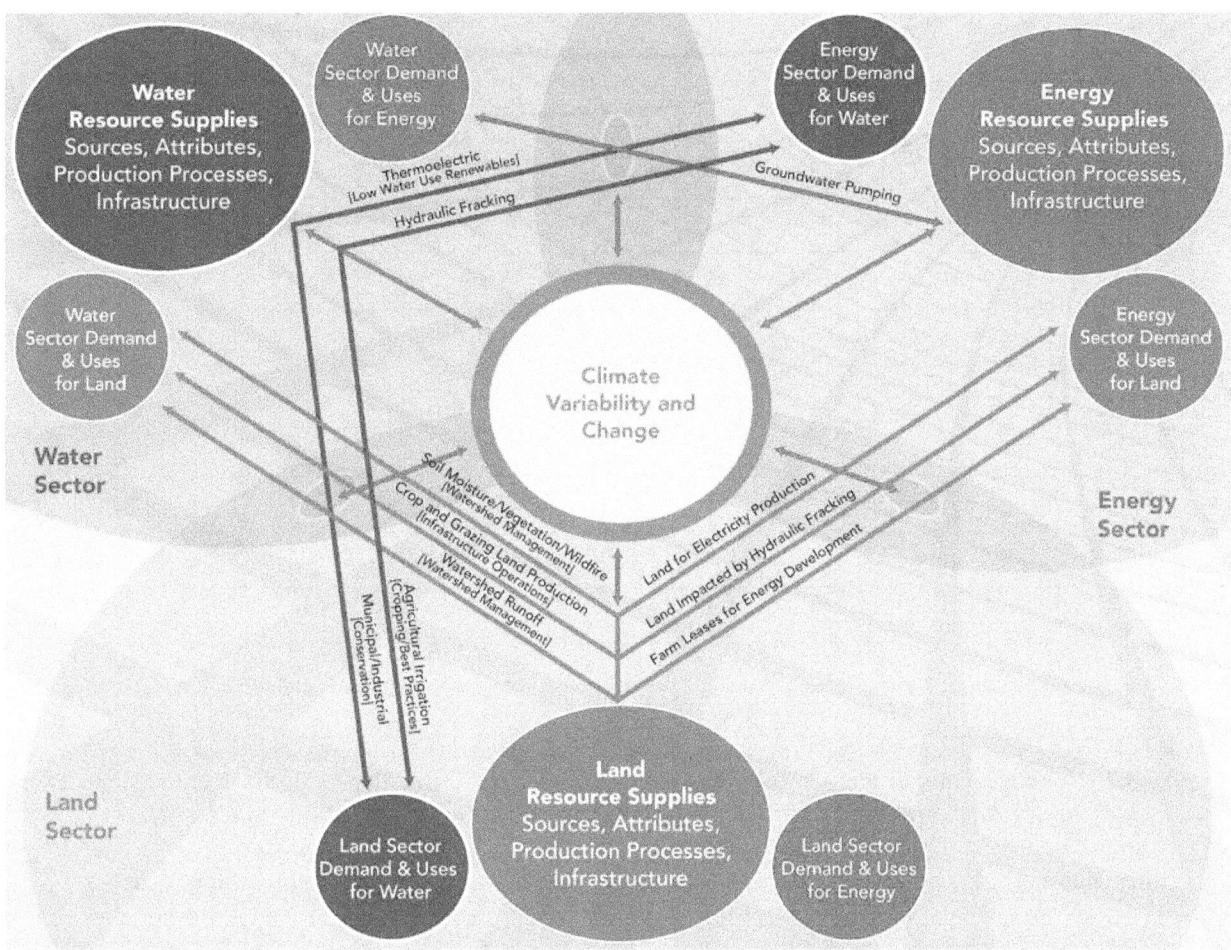

Figure 3.1. Interdependent interfacial linkages for the EWL nexus specific to the 2011 Texas drought and heat wave. Key interfacial interactions and associated mitigation/adaptation options (in brackets) are graphed in the context of our "resource demand, supply endowment, and technologies" conceptual model (Figure 2.3). See sections 3.2.2 and 3.2.3 for a description of the specific interfacial interactions.

3.2.1   Background on Texas Drought/Heat Wave

While parts of New Mexico, Oklahoma, Nebraska, and Louisiana experienced extreme to exceptional drought in 2011, Texas was the epicenter of the event, with the entire state experiencing drought. However, the Texas drought was unique beyond its geographic extent. As Figure 3.2 shows, Texas experienced both the hottest and the driest conditions on record. In fact, the 2011 summer was over 2.5 °F hotter and 2.5 inches of rain drier than previous record highs (1980 for temperature and 1956 for precipitation). The drought's severity appears to be connected to natural variability, specifically the occurrence of La Nina events in the Pacific Ocean, as well as the Pacific Decadal Oscillation (PDO) and Atlantic Multidecadal Oscillation (AMO). "The record warm weather during the summer in Texas was primarily a consequence of the lack of rainfall" (Nielsen-Gammon 2011).

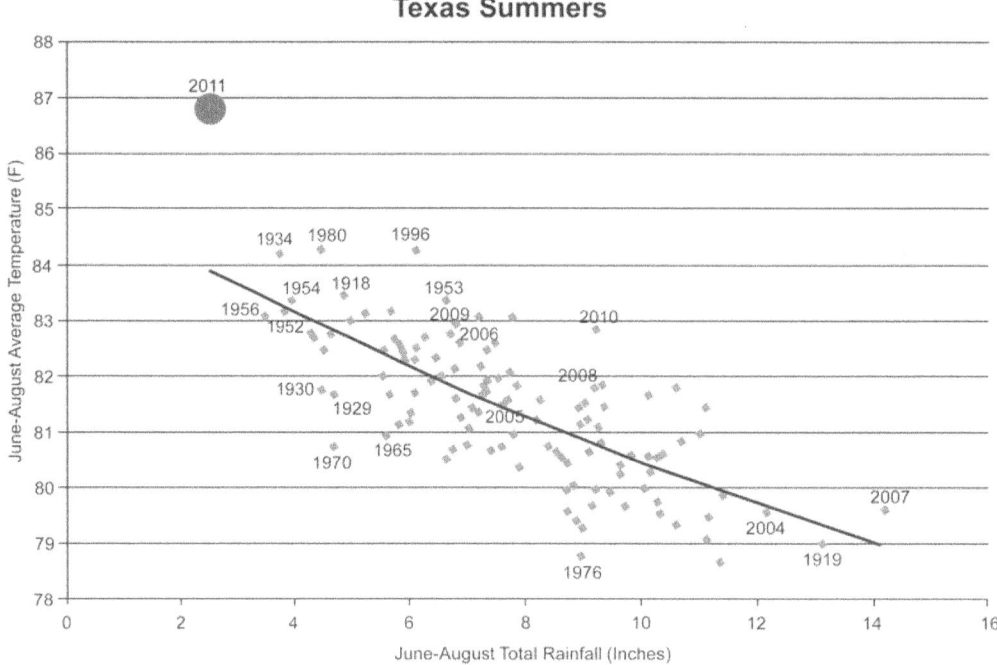

Figure 3.2. Average summer temperatures and rainfall for Texas. Note that both the average summer temperature and rainfall for 2011 were outside the range of variability since 1918.

Other measures also point to the severity of the 2011 drought. According to an article in *Texas Climate News*, Texas' average Palmer Drought Severity Index (PDSI; Palmer, 1965) in the summer of 2011 (June through August) was -5.37, the lowest (indicating the most severe drought conditions) since the instrumental record started in 1895 (Dawson 2012). The article further states:

> According to the federal government's National Climatic Data Center [sic estimates based on tree-ring analyses], "there was apparently only one other year during the last 461 years when Texas had a drought so severe. When this error band [for drought estimates via tree ring information] is taken into account, there is only one value in the paleo-record, 1789 (-5.14), that can be said to be equivalent to the 2011 observed value. Thus, 2011 appears to be unusual even in the context of the multi-century tree-ring record."

3.2.2 Competition Among and Between Energy-Water and Land-Water Interfaces

Competition for water is the most apparent process linking energy, water, and land (Figure 3.1). In 2004 water withdrawals in Texas equaled 0.4 million acre-feet (MAF) for thermoelectric production, while land-related withdrawals were 3.9 MAF for municipal, 1.7 MAF for industrial, and 9.5 MAF for agriculture (Texas Water Development Board 2011). However, the 2011 drought/heat wave, which saw both above average temperatures and less precipitation, depleted critical stocks of water (soil moisture, reservoir volumes, and stream flow) while increasing demand across most use sectors. As a result, there was not enough water to meet the basic needs described above. This caused direct competition between energy-based and land-based water demands, not only between the energy and land interfaces but within these sectors as well.

The 2011 drought and heat wave depleted much of the reservoir storage across the state, which is currently threatening thermoelectric generation. "More than 11,000 megawatts of Texas power generation — about 16 percent of the Electric Reliability Council of Texas' (ERCOT) total power resources — rely on cooling water from sources at historically low levels. If Texas does not receive "significant" rainfall by May, 2012, more than 3,000 megawatts of this capacity could be unavailable due to a lack of water for cooling" (ERCOT 2011). This potential impact is further intensified when considering that increased cooling demands accompany summer temperatures. In the case of the 2011 heat wave, peak electricity demands soared to all-time highs, exceeding the prior record on 8 of the first 12 days of August 2011. The peak demand rose to 68,294 MW closely approaching the state's capacity of 72,000 MW. During this time wholesale prices rose to as much as 60 times normal summer prices (Smith 2011). While the state's growing utilization of wind power (Texas is the state with the largest wind power generation capacity at 10,223 MW (AWEA 2012)), currently 12.5% of the state's production, reduces challenges over limited water supplies it places the state at greater risk of meeting peaking demands due to the inherent variability of production.

Competition over water occurs across multiple energy-water bilateral interfaces. Beyond thermoelectric generation, water scarcity is also threatening shale gas production. In 2010, the Texas Water Development Board estimated that 13.5 billion gallons of water were used in the drilling and stimulation of gas shale wells in Texas. Because of intensifying drought conditions, and competition with other water use sectors, energy developers are finding it increasingly difficult and/or costly to obtain water (see Illustration 3.1). While natural gas prices are influenced by many factors, reduced production would put upward pressure on the price. Relatively small changes in gas prices could favor electricity production by coal, regionally increasing water demands (see Macknick et al. 2011) and greenhouse gas (GHG) emissions.

Land-based water demands have also faced challenges induced by the drought. According to the Texas Commission on Environmental Quality, thirteen communities are on their "high priority" water list, meaning they could run out of water within 6 months, or do not know how much water they have remaining. In fact, one community recently ran out of water requiring water be trucked in from a source 10 miles away (Fernandez 2012). These shortfalls are occurring while most Texans are adapting to water shortfalls through restrictions that limit or prohibit outdoor watering of lawns, washing cars, and other "discretionary" water uses. Water for irrigation is also in limited supply, as is evident in South Texas where farmers have been warned of potential need for rationing in 2012 (Ana 2011).

Drought combined with strong population and economic growth has encouraged water users in the state to consider various strategies to adapt to a water-limited future. Contrasting local examples of expanded water transfers are described next. Sweetwater, a town in West Texas, is moving ahead with a plan to sell 250 million gallons of water to a proposed coal-fired power plant being built in the region despite environmentalist claims that the dry region does not have enough water to sell (Plushnick-Masti 2011). Alternatively, Stillwell et al. (2011) suggest that thermoelectric power plants could provide a future source of water for municipal and industrial growth in Texas. Specifically, the proceeds from selling water rights from thermoelectric power could be used to retrofit the plant with hybrid or dry cooling (dry cooling relies solely on air to cool the working fluid that drives the electric generators in thermoelectric power plants; hybrid uses a mix of water and air), reducing or eliminating water use. In fact, public opinion is driving most new thermoelectric development to consider dry cooling—examples include the Texas Clean Energy Project near Odessa (coal gasification for CO_2-EOR, ammonia production, and electricity), Tenaska Trailblazer near Abilene (post-combustion CO_2 capture coal-fired power plant for CO_2-EOR), and White Stallion in Matagorda County (proposed bituminous coal-fired power plant looking to use a non-potable water source).

A potential unintended consequence of both the planned water transfers and expanded use of dry cooling is increased demand for electricity. Should the transfers require pumping, additional electricity would be required. Likewise, power plants using dry cooling experience efficiency losses on hot days (Maulbetsch and DiFilippo 2006), thus decreasing electricity generation. Ultimately these increases in electricity demands could shift water demands to other areas of the state (i.e., where the additional electricity is generated).

Illustration 3.1: Climate-Related Decision Making at the Nexus
Switching from Coal to Natural Gas: The Role of Hydraulic Fracturing

The last 10 years have seen a 50% increase in electricity generation by natural gas fueled power plants (EIA 2011). Recent production has lowered the cost of natural gas, driving further expansion of the domestic natural gas industry. Between 2006 and 2010, the US production of natural gas, particularly from shale gas, has increased 6-fold: from 2.7 billion cubic feet per (bcf) per day to an estimated 13.3 bcf per day (Newell 2011). Most of the natural gas extracted in the United States comes from the Marcellus Shale (Northeast), the Gulf Coast, and the Barnett Shale and Woodford Complexes (Mid-West and Texas region) (Figure 3.3a). New technologies are allowing access to previously unavailable sources (e.g. Marcellus), and extraction activities in the West, particularly in the Bakken Shale (Montana, North Dakota) are increasing in response to increasing demand, and there are other regions with prospective natural gas reserves that may be tapped (Figure 3.3b).

Greenhouse emissions from natural gas are about half those associated with coal (although a comprehensive end-to-end assessment and comparison is lacking). Consequently, natural gas can be expected to be a part of any GHG mitigation strategy. Projections indicate that natural gas use for transportation fuels and electricity generation in the United States will continue to grow through 2035 (EIA 2011). But there are impacts particularly to water systems related to natural gas extraction. Natural gas is extracted from shale using a process known as hydraulic fracturing. The process uses a high-pressure fluid (water and a mix of chemicals) to create and widen fissures in rock to access natural gas. For each well, 3-4 million gallons of water is required over the lifetime of the site, depending on the geologic structure being accessed (DOE 2009). There are also unclear impacts on the quality of groundwater resources (EPA 2011b).

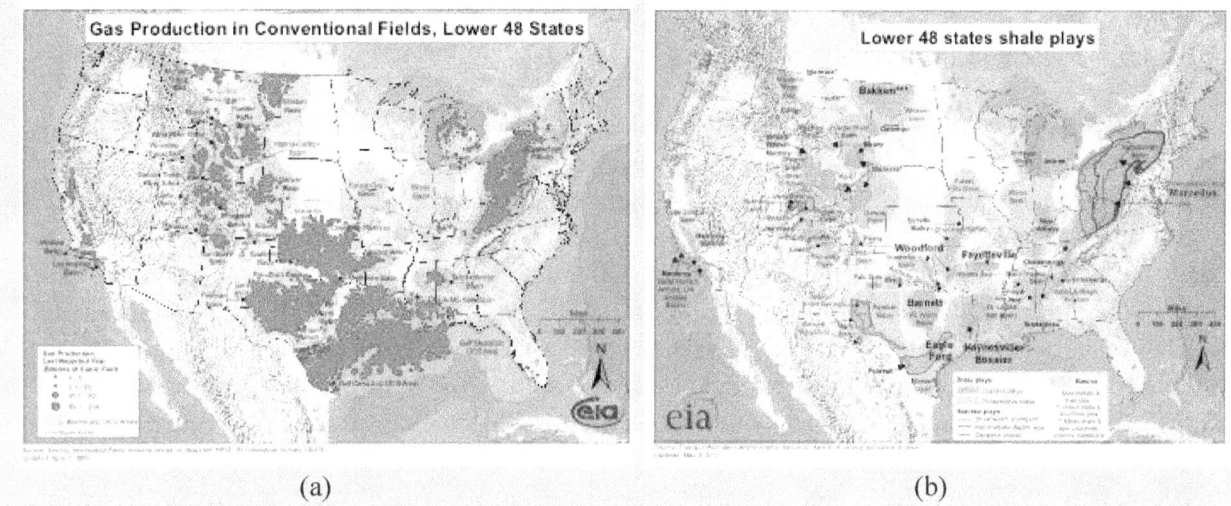

| (a) | (b) |

Figure 3.3. (a) Natural gas production and (b) shale gas plays in the coterminous U.S. Sources: EIA 2008, EIA 2011.

Consideration in Decision Making:

A carbon constraint or carbon tax would be expected to further increase the fraction of natural gas in a national fuel portfolio. In some places, power plants are already switching from coal to natural gas, thereby increasing demand for natural gas resources. In Colorado, the Clean Air-Clean Jobs Act passed in 2011 is prompting coal-fired power plants in the region to switch to natural gas or renewable technologies by 2017, which will reduce relative GHG emissions. This would require an increase in natural gas production. However, the implications for water resources are ill-constrained.

In Texas, however, lack of water for hydraulic fracturing of shale gas reservoirs has hindered natural gas production. According to an October 2011 article in Stateline (Malewitz 2011):

> "Fracking has stirred controversy in towns across the country largely because of fear that it contaminates drinking water. But in Texas, where the vast majority of oil wells have been fracked, it's the industry's water-sucking tendency that has become a divisive issue. In August the town of Grand Prairie, in the northern part of the state, became the first in Texas to enact a ban on the use of water for fracking.

> To address the water shortage driven the 2011 drought in Texas, companies have trucked water across city lines. Others have added infrastructure that enables recycling or reduced water use, but at significant cost. Still others have purchased water from farmers—an option that becomes less viable as drought persists."

3.2.3   Cascading Impacts on Water-Land Interfaces and Implications for Energy Production

Beyond the impacts on water supply, the 2011 drought has also changed the vegetative landscape of Texas (Figure 3.1). Measured changes have been documented in terms of cropping, grazing, and wildfire damage. These changes tend to reinforce and intensify individual impacts on land and water resources.

To a lesser extent, these changes feedback through water and land use to impact energy demand and production.

The Texas drought/heat wave demonstrates multiple linkages between climate and the water-land interface: specifically, higher temperatures and reduced rainfall have devastated both rain-fed agriculture (leading to abandoned cultivation) and irrigated agriculture (causing reduced yields). The Texas AgriLife Extension Service recently estimated the drought cost Texas farmers $3.1 billion due to reduced yields coupled with the farmers' inability to take advantage of unusually high commodity prices. Losses were greatest for cotton, followed by hay, corn, wheat, and sorghum. Not considered were losses to fruit and vegetables, horticulture, nursery, and other grain and row crops (Fannin 2011). "One of the most telling aspects of the 2011 drought was that irrigated farms were not spared. While most irrigation systems in Texas work well in normal or even below normal rainfall conditions, many irrigators found that water supplies could not meet the requirements of the crop without any rain and with the excessive heat. By mid-July, farmers began to try to stop economic losses by using all of their water supplies for fewer acres as water demand exceeded supply" (Fannin 2011).

Additional water-land linkages are evidenced by the connection between rangeland, livestock, and feed production. The AgriLife Extension Service estimated losses of $2.06 billion for livestock due to the 2011 drought/heat wave (Fannin 2011). These losses were in part due to the drought-decimated rangelands resulting in smaller stock to market and/or farmers selling off their stock (flooding the market and thus reducing prices). In fact, 84% Texas and Southwestern Cattle Raisers Association members surveyed had recently reduced their herd, with an average reduction of 38% (White 2011). Likewise, reduced crop yields, as noted above, led to increased feed costs. In total, the $5.2 billion in losses represent 27.7% of the average value of agricultural production over the last four years (Fannin 2011).

Drought has also impacted forests and grasslands, with downstream implications for water production. An estimated 100 million to 500 million trees with a diameter of 5 inches or larger on forestland were estimated to have succumbed to the drought. That range is equivalent to 2 to 10% of the state's 4.9 billion trees (Texas Forest Service 2011a). The resulting high fuel loads, low water content, and high temperatures led to a record wildfire season. From Nov. 15, 2010 through Sept. 29, 2011 Teas saw 23, 835 fires that destroyed 2,763 homes burned a record 3.8 million acres—an area about the size of Connecticut (Texas Forest Service 2011b).

The combined drought-induced changes in cultivation, irrigation, grazing, and wildfire represent significant land use changes over a very short period. Changes in land use will impact soil moisture and albedo, potentially intensifying drought through increased ground temperatures and modification of local moisture cycling. The depleted stocks of soil moisture will significantly affect watershed runoff, prolonging reduced stream flow and reservoir storage.

These land-water interfacial changes could affect energy services. Reduced surface water and soil moisture supplies may encourage increased groundwater pumping to meet irrigation demands. More pumping combined with reduced recharge would result in greater pumping depths and hence increased energy demand. Increased erosion could lead to higher sediment concentrations and thus increased treatment burden to municipal water systems. Finally, the expanding footprint of wildfire increases the risk of disruption to the energy distribution infrastructure (transmission lines and pipelines).

Rural family farms provide another example of climate, energy, land, and water interfaces converging in Texas. In today's economy, it is becoming increasingly common for family farms to depend on multiple streams of revenue, including ranching, farming, mineral rights and/or development of solar and wind resources. Increasing market volatility and uncertainty about the climate is making it increasingly difficult for landowners to manage limited water resources for both energy and agriculture. Management choices will have important implications for both future land and mineral rights utilization.

Although a bit overwhelmed by the severity of the current drought/heat wave, Texas has drought programs in place to deal with this extreme event and to adapt to future events (National Drought Policy Commission 2004). Adaptation programs are generally triggered by a particular event, such as declining groundwater levels, reduced stream flow, low reservoir levels, or low PDSI. The first response to be triggered is the assembly of a task force to coordinate response across the various state agencies and liaison with federal response. The task force is also responsible for monitoring, reporting, and communicating information about the drought and its impacts. Other support is triggered in the form of fire prevention teams and suppression forces to support local fire departments; potential evapotranspiration networks coupled with crop models to accurately estimate irrigation needs; teams to monitor and supply potable water where municipal systems fail; and community outreach programs that encourage water conservation.

Other drought programs are aimed at recovery and adaptation to future extreme events. Support includes assistance through soil and water conservation districts to develop soil and water conservation plans and drought contingency plans; education of land owners in rangeland best practices for brush and grazing management; drought recovery assistance (low interest loans) to implement water/soil conservation strategies for both land owners and municipalities; demonstration projects of proper irrigation practices for athletic fields; and monitoring coupled with forest/aquatic/faunal management by the Texas Parks and Wildlife Department (National Drought Policy Commission 2004). General water planning by the Texas Water Development Board is also a key component of any future adaptation planning (Texas Water Development Board 2012). Ultimately, the adaptations made following the current drought/heat wave will dictate future dynamics relating to extreme climate events and the impacts on energy, water, and land resources.

3.2.4   Projected Changes in Competition for Land and Water

Drought is certainly not unique to Texas, nor is the potential for climate change to intensify drought. Figure 3.4 shows the projections of the mean changes in the extreme PDSI for the 30-year period centered on 2050. The results—from the average of 22 climate models—are based on an intermediate emissions scenario, A1B, from the Fourth Assessment Report of the Intergovernmental Panel on Climate Change (see Strzepek et al. 2010 for more information). The magnitude of any specific drought event could be much stronger, as Figure 3.3 only conveys projected means. Although the current drought in Texas provides a useful analogy for other regions, important differences exist. For example, hydropower production, which is of minor importance in Texas, is vulnerable to drought both through reduced reservoir storage and through modified reservoir operations (National Energy Education Development Project 2007). Lost production can affect both base and peaking load, but possibly most important is loss of flexibility to use hydropower to smooth variability in electricity generated by PV and wind sources. While thermoelectric production is challenged by limited water supply in Texas, other regions, particularly in the east, are challenged by thermal limits on cooling water effluent (Averyt et al. 2011;

Union of Concerned Scientists 2007). Another important difference is in regional competition over limited water supplies. West of the 100th meridian (Figure 3.4), where precipitation is largely less than 20 inches per year, significant demands are placed on water resources by agriculture (Kenny et al. 2009). In contrast thermoelectric water use dominates east of this boundary (except in the lower Mississippi Valley and parts of Florida) where significant water supplies are available for open loop cooling (where stream or lake water is directed through the power plant and across the condenser coils to cool the working fluid and then discharged directly back to the source water body). Drought impacts on electricity demand may be limited in some northern regions where homes and businesses lack electric cooling.

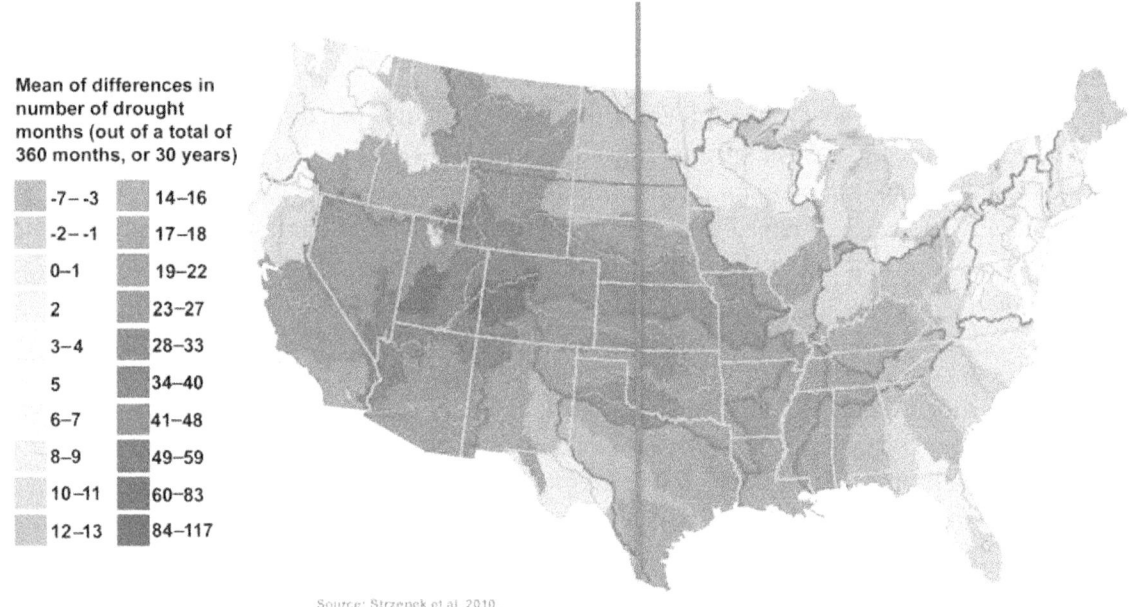

Mean of differences in number of drought months (out of a total of 360 months, or 30 years)

-7– -3	14–16
-2– -1	17–18
0–1	19–22
2	23–27
3–4	28–33
5	34–40
6–7	41–48
8–9	49–59
10–11	60–83
12–13	84–117

Source: Strzepek et al. 2010.

Figure 3.4. Projections of the mean changes in the extreme Palmer Drought Severity Index (PDSI) for the 30-year period centered on 2050. (Figure is from Averyt et al. 2011 redrawn from Strezepek et al 2010). Approximate boundary of 100th meridian represented by blue line.

3.2.5 Summary of 2011 Texas Drought and Heat Wave Impacts on EWL Interfaces

The 2011 Texas drought and heat wave provides insight into future extreme climate events and their linkages to energy, water and land resources. Additionally, observation and analysis of system interactions provide indications about the resilience, adaptive capacity, and potential risks. The 2011 drought and heat wave was seen to intensify competition over limited water resources; specially, water for energy production (thermoelectric generation and natural gas well fracking) and land-based demands (e.g., farming, ranching, municipal purposes) (Figure 3.1). High temperatures exacerbated competition by increasing demand of water for crop/lawn irrigation and electricity generation to meeting increased use of air conditioning. At the same time the ability of Texas watersheds to produce water was compromised by changing vegetation, cropping, grazing, and wildfire conditions. These watershed changes were seen to reinforce and intensify individual impacts on land and water resources (e.g., reduced cropping raises feed prices that change grazing patterns that impact vegetation density and thus impact wildfire vulnerability). These same watershed changes impacted the energy sector through increased electricity demand for groundwater pumping and municipal treatment of waters fouled by watershed erosion, while putting transmission infrastructure at risk by wildfire danger.

Ultimately, adaptive measures taken today and recovery/planning efforts following the 2011 drought and heat wave will dictate future dynamics in Texas relating extreme climate events and the impacts across the EWL interfaces. Measures such as adopting low water use renewables for electric generation, use of non-potable water for gas well fracking, municipal/irrigation conservation, transfers of water rights and watershed management will both reduce water demand while improving water production. Such measures would have the combined effect of restructuring the competition over water in both time and space.

3.3 Regional Differences

Various regions across the U.S. are characterized by different physiographies, economies and availability of natural resources, and thus each region will respond to climate variability and change in very different ways. This results in in different dynamics between the EWL interfaces. Because of the importance of regionalization, this impact is explored in some detail here. The focus of this comparison is California and the Gulf states.

We begin by reviewing potential climate change futures for both regions. We then consider several key themes that distinguish these regions in terms of the interfacial dynamics forced by climate variability and change as well as differences in institutional and adaptive capacity. The key interfacial interactions and accompanying mitigation/adaptation options are graphed in Figure 3.5 for California and Figure 3.6 for the Gulf states.

3.3.1 Climate Projections

California is characterized by a complex geography with a broad mix of climate zones, including alpine, arid, semiarid, Mediterranean and marine. Although climate projections vary across these zones, some general trends are evident. By century's end, mean annual temperatures are projected to increase by 5-7°F above recent historical levels (1970–2000) under the Special Report on Emissions Scenarios (SRES) A1 scenario and 2-4°F under the SRESB2 scenario (NCA 2013b). This will increase evapotranspiration rates and drive earlier snowmelt runoff. General precipitation trends are much more variable, with the SERSA1 scenario suggesting relatively little change in mean annual precipitation in the north and a 15–20% decrease in the south, while the SERSB1 scenario projects that most of the state will experience a 10–20% decrease. In general, winter precipitation is expected to increase slightly, while decreasing precipitation is expected for the remaining three seasons. Extreme events are likely to intensify in terms of drought, heat waves, and floods. Probably most important is the projection that drought will increase in frequency, magnitude and duration. Together, these projected changes would significantly affect water resources. Recent studies have suggested reductions in Colorado River flows from less than 10% to about 50% by mid-century due to climate change (Christensen and Lettenmaier 2007). Sea level rise of 7-55 inches is also projected.

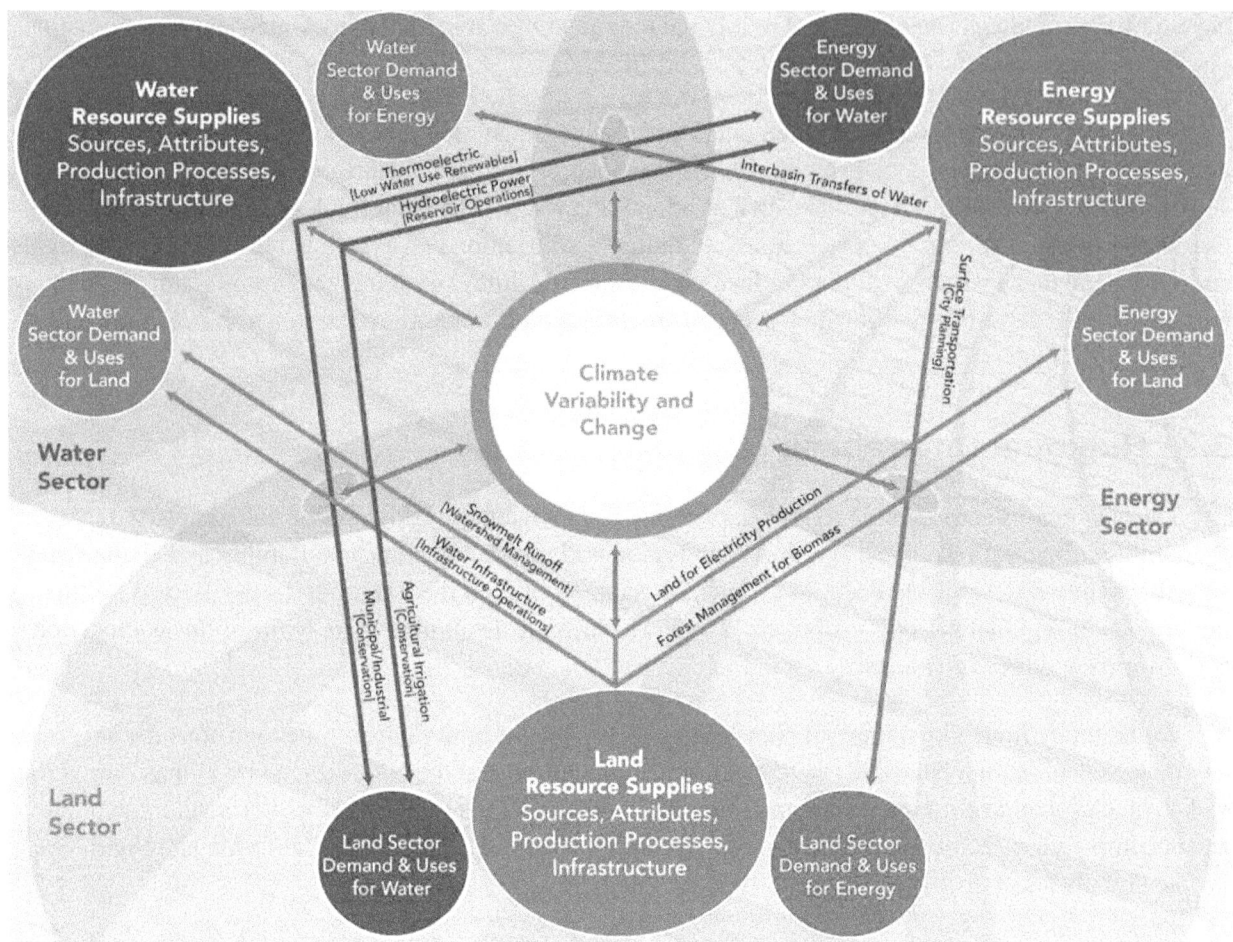

Figure 3.5. Interdependent interfacial linkages for the EWL nexus specific to the California region. Key interfacial interactions and associated mitigation/adaptation options (in brackets) are graphed in the context of our "resource demand, supply endowment, and technologies" conceptual model (Figure 2.3). See section 3.3 for a description of the specific interfacial interactions.

The southeast Gulf climate differs markedly from California and the rest of the U.S. The climate is uniquely warm and wet, with mild winters and high humidity (Karl et al. 2009). Climate models project a general warming trend across the region: 4.5 °F by the 2080s under low emission scenarios ranging up to 9 °F under high emissions (Karl et al. 2009). The number of very hot days will increase at an even faster rate than temperature. Although precipitation projections are mixed (for example, some models predict more precipitation for south Florida, and some less), the frequency, duration, and intensity of droughts are likely to increase (Karl et al. 2009, UCS 2012).

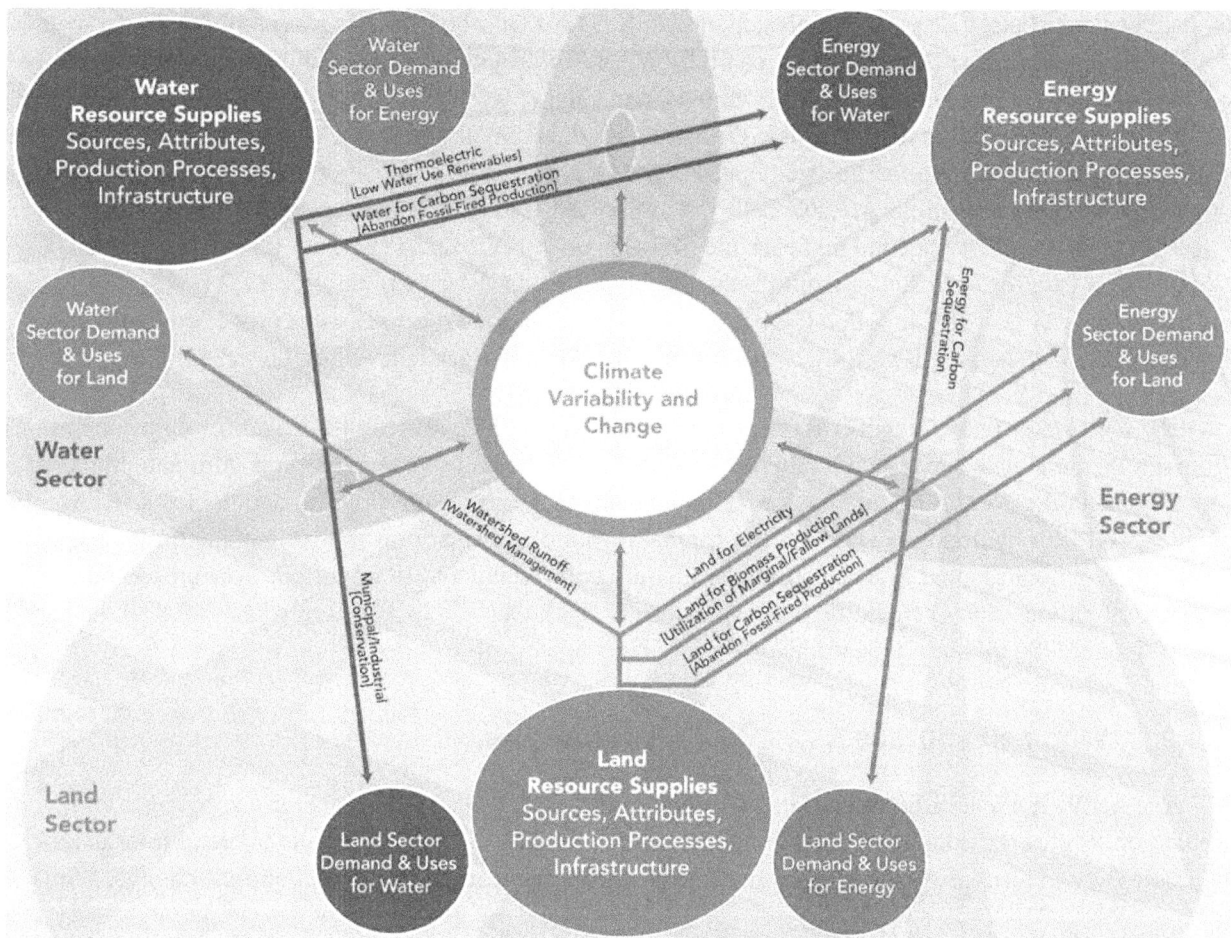

Figure 3.6. Interdependent interfacial linkages for the EWL nexus specific to the Gulf states region. Key interfacial interactions and associated mitigation/adaptation options (in brackets) are graphed in the context of our "resource demand, supply endowment, and technologies" conceptual model (Figure 2.3). See section 3.3 for a description of the specific interfacial interactions.

The Gulf states are particularly vulnerable to sea level rise and increased hurricane intensity. Sea levels are likely to rise between 15 and 40 inches along the coast by 2100 (UCS 2012), with the region being particularly sensitive due to subsidence and general low-lying land. Although precipitation projections are unclear, hurricanes are expected to intensify, producing extreme rainfall events.

3.3.2   Energy Portfolio

California and the Gulf Coast states differ in how energy is produced and consumed, particularly in terms of electricity generation. Differences in the generation portfolio have important implications for greenhouse gas emissions, water use, and future adaptive capacity. These differences, along with the future evolutionary path each region's portfolio follows will dictate the dynamics and linkages between climate, energy, water, and land.

According to 2007 plant-by-plant data (EPA 2011a), California generated approximately 24 GW of electricity each year (24 GWyr) from an installed capacity of 67 GW. Most is generated through natural gas (55.6%), nuclear (16.0%), and hydroelectric (13.2%) power plants. California is defined by a lack of in-state coal-fired electricity: six plants generate only 1% of California's electricity. In 2007, just over 25% of all in-state electricity generation came from renewable sources, including hydro (13.2%), geothermal (6.1%), biomass (2.7%), wind (2.6%), and solar (<0.3%). Excluding hydropower, California produces more than double the renewable electricity produced in Texas (2.8 GWyr versus 1.2 GWyr), the second highest ranked state for electricity production. California also imports more electricity than any other U.S. state. For example, in 2007 California imported 10.6 GWyr, more than double the electricity imported by Virginia, the second highest state (4.7 GWyr) (EPA 2011a). Further, California Energy Commission data for 2008 shows that the state generated 23.8 GWyr and imported 12.2 GWyr, approximately 34% of consumption (Nyberg 2009). This imported electricity is a mix of coal (18.2%), large hydro (11.0%), natural gas (45.7%), nuclear (14.5%), and other renewables (10.6%).

The Gulf states (excluding Texas) generated 76 GWyr in 2007 (roughly triple California's generation) from a total nameplate capacity of 182 GW (EPA 2011a). The region's generation is dominated by coal (45.2%), natural gas (28.6%), and nuclear (19.4%). Only 4.1% of electricity is generated from renewable sources, 1.1% for hydropower and 3.0% for biomass. Wind and solar power are almost entirely absent. This region generates most of its electricity, with only approximately 3% imported. Alabama exports 29.3% of their electricity generation, while Florida (12.3%), Georgia (5.4%), Louisiana (29.8%), and Mississippi (7.1%) all import electricity.

3.3.3   Water Resources

Another theme that differentiates the two regions is water resources, both in terms of supply and utilization (Figure 3.5 and 3.6). These fundamental differences give rise to unique bilateral interfacial dynamics that in turn interact, compete, and evolve very differently in the face of climate change.

California's water supply depends heavily on high elevation precipitation, particularly late spring snowmelt runoff and an extensive reservoir system. The principal sources of this water, the Sierra Nevada and the southern Rockies (in the form of flows of the Colorado River), are geographically displaced from the centers of demand. As a result, California has constructed a sophisticated system of reservoirs and conveyance structures (e.g., California Aqueduct, Colorado River Aqueduct, Los Angeles Aqueduct) to transport water to the central valley for irrigation and to the coast for municipal needs. This extensive canal structure has resulted in a significant energy "mortgage" for the state to lift and move the water. Specifically, California uses 7.7% of its total electricity to capture and move water to its customers (California Public Utilities Commission 2011).

Like much of the Southwest, the water resources available to California are largely appropriated (rights for a particular use of the water have been assigned), and have been over-appropriated in some cases, like the Lower Colorado River (California Department of Water Resources 2009). Because of this heavy use of available water, the state closely administers water rights in accordance with the principle of "prior appropriations" (i.e., the first in time to make beneficial use of water has first priority to limited water supplies in times of drought), which is very different from the means of water management found in the eastern U.S. Finally, water demands in California are structured very differently from the Southeast (Table 3.1); specifically, irrigated agriculture dominates withdrawals (76.7%), while the municipal sector is responsible for 22.7% of the withdrawals, industrial 0.4%, and thermoelectric generation 0.3% (Kenny et al. 2009).

Climate change threatens further stress on a system that is already struggling over competing demands for limited water resources. Higher temperatures will increase evaporation and decrease soil moisture, affecting reservoir storage and watershed production. Higher temperatures also mean earlier snowmelt runoff. Coupling this with intensifying storms could make it hard to balance reservoir storage with flood control. Finally, current demands will be intensified by higher temperatures because more water will be needed for crops, landscaping and the electricity demanded for home and commercial air conditioning (California Department of Water Resources 2009).

The projected stresses coupled with a system in which the water resources are largely appropriated suggest that California will have little fresh water available for new development (California Department of Water Resources 2009). Under such circumstances, new development requires the transfer of a water right from an existing use (e.g., retirement of use) or use of a non-fresh water source. Ultimately, this creates competition for limited resources; specifically, competition for water for electricity generation (hydroelectric and thermoelectric), extraction and refining of primary fuels, and irrigation, municipalities, and the environment. How this competition plays out will strongly influence the evolution of the energy, agricultural, and municipal sectors; that is, the future portfolio of electric power plants, transportation fuel choices, land use, and agricultural production. California has helped ease this competition by requiring new thermoelectric power plants to fully exhaust alternative water sources before considering freshwater (California Water Code, Section 13552).

The Gulf states have a very different set of water issues than California. The region's surface water supplies primarily rely on rain-fed (as opposed to snow-derived) water that has historically been distributed rather uniformly across any given annual cycle. Changes in precipitation due to climate change are expected to be moderate (Karl et al. 2009), and whether precipitation increases or decreases, climate will affect water resources less in the Gulf states than in California. Consequently, extreme events, particularly flooding and possibly drought duration, are likely to have more influence than long-term precipitation changes.

Gross withdrawal amounts for California and the Gulf states are remarkably similar. For instance, in 2005, California (then population 36 million) withdrew 45.7 billion gallons of water each day (45.7B gal/d) while the Gulf states (population 38 million) withdrew 49.3B gal/d (Table 3.1). The breakdown between fresh and saline waters was also similar: 32.9B gal/d fresh and 12.9 B gal/d saline water for California, and 37.6B gal/d and 11.7B gal/d respectively for the Gulf states. Even withdrawals for groundwater and surface water are comparable: 11.0 B gal/d (groundwater) and 34.8B gal/d (surface water) for California compared with 9.8B gal/d and 38.4B gal/d for the Gulf states (Kenny et al. 2009).

Table 3.1. Water withdrawals for California and the four Gulf states, in million gallons per day (adapted from Kenny et al. 2009).

State	Domestic		Irrigation	Livestock	Aquaculture	Industrial		Mining		Thermoelectric power		Total		
	Public Supply	Fresh	Fresh	Fresh	Fresh	Fresh	Saline	Fresh	Saline	Fresh	Saline	Fresh	Saline	Total
CA	6,990	486	24,400	197	646	72	23	53	255	50	12,600	32,900	12,900	45,700
Gulf	5,778	471	5,219	143	506	4,574	24	725	151	20,238	11,537	37,600	11,711	49,340

Although gross withdrawals are comparable, California and the Gulf states have a significantly different water use patterns. For example, California uses only 20 million gallons a day (0.02B gal/d) of fresh water for thermoelectric power cooling, whereas the Gulf states withdraw 20.2B gal/d (Table 3.1). Principally, this is because California uses freshwater for closed loop cooling whereas the ready availability of surface water in the Gulf states allows for once-through cooling. Water consumption rates for thermoelectric power generation are similar in both regions.

The two regions also significantly differ in water use for agriculture and industry. For instance, California withdrew approximately 24.4B gal/d for irrigation compared with only 5B gal/d for the Gulf states. And California only withdrew 0.07B gal/day for industry compared with 4.6B gal/d in the Gulf states. Both regions withdraw comparable amounts for residential consumption.

Although the Gulf states have much more abundant water supplies, with enough for future development (outside of Florida), future climate change will likely stress the energy-water-land interface considerably. For example, Georgia has already had significant droughts in the last decade, putting significant strain on the thermoelectric generation system. Elevated water temperatures from water returned by thermo-electric plants to streams have also created tension among different uses. For example, the Browns Ferry nuclear power plant in Alabama had to shut down in the summers of 2010 and 2011 due to the ambient river water temperature being above 90°F (Energy Collective 2011); discharged water from the power plant cannot exceed 86.9°F. These impacts are expected to degrade over the remainder of the 21st century. A likely response is that new power plants in the Gulf states will use closed loop cooling (or evaporative cooling withdrawals a volume of water for cooling which is continuously recirculated until it is completely evaporated as such this means of cooling uses a much smaller amount of water than open-loop cooling, but it is all consumed). Closed loop cooling systems are prevalent in the U.S. west, where river flows are generally much lower. However, closed loop cooling provides a couple of key advantages for the U.S. southeast: water is not discharged back to the river and therefore discharge rules are circumvented, and closed loop plants are more resilient to drought since they do not necessarily require large river flows. The Gulf states are also likely to be affected by other water quality issues, including the potential onset of industrial scale biomass and biofuels. For instance, the Gulf of Mexico is already annually affected by a massive algal bloom due to excessive nutrient loading in the Mississippi and other rivers. The formation of this bloom is likely to be exacerbated through both climate change and further nutrient runoffs from widespread biomass growth for biofuels.

Water is accessed differently in the Southeast in terms of water rights. Whereas California is administered by prior appropriations, the Gulf states are largely controlled by the riparian doctrine (Mississippi is not). These states developed the doctrine based on English Law, and allow anyone with land bordering a water source to make reasonable use of the water. Typically, water rights are exclusively linked to land ownership and water cannot be transferred out of the source watershed (i.e., no inter-basin transfers); this is very different from the prior appropriations doctrine in the West. During water shortages, rights are allotted proportionally to frontage on the water source.

3.3.4 Extreme Events

Climate change poses a wide range of threats to U.S. energy-water-land resources. These vulnerabilities threat differ regionally based on the physical, ecological, social, economic, and other features unique to each region.

Although California is subject to a range of extreme events, the state is particularly vulnerable to drought and heat waves. This vulnerability is largely a function of the limited availability of water and the high temperatures of the state's desert and semi-arid regions. Climate projections suggest drought will become more frequent, more severe, and last longer; this is coupled with rising temperatures. California is no stranger to drought, recently enduring mean annual stream flows below 65% of normal for three consecutive years (2007–2009) (California Department of Water Resources 2009). Because we have already discussed several of the dynamics linking drought and the EWL nexus as part of the 2011 Texas drought, here we focus on several linkages unique to or pronounced in California.

The structure of water demands in California, with irrigated agriculture the dominant withdrawal, creates a strong linkage with drought. While competition among energy, municipalities, and agriculture remain, the large disparity in use offers an opportunity for short-term leases from low value agriculture (e.g., irrigated pasture or alfalfa) to higher valued agricultural, municipal, and thermoelectric users. In other words, low value agriculture is available to buffer the sting of drought.

Another important factor is the impact of drought on hydroelectric production in California. Since 1990 hydroelectric power has satisfied about 11% of the region's total electricity demand, with year-to-year variations ranging from 8–16% (EIA 2011). Several studies have explored potential climate impacts on hydropower production, with projections ranging from a slight increase to a 60% decrease in production (Harou et al. 2010; Georgakakos et al. 2011; Vicuna et al. 2008). Because of the broad western states electric interconnection (i.e., Western Electric Coordinating Council), the ultimate effect of lost hydropower production in California depends also on conditions in the Columbia River and the Colorado River basins (Cayan 2003). Cross-basin linkages can compensate for hydropower losses when at least one of the basins experiences wet conditions or intensifying effects when two or more are dry. Beyond the obvious loss of electricity generation, loss of hydropower affects the ability of interconnection managers to balance intermittent loads from other renewable generators. This cascading loss of electricity production puts a greater burden of responsibility on thermoelectric generation and thus downstream demand for water. Ultimately, this all results in a complex interplay of water competition between energy and land resources.

Warm summer temperatures characteristic of much of California and which are projected to increase with climate change, pose a compounding risk to electric service due to feedback across multiple land and water interfaces. The difficulty begins with increasing peak demand for home and commercial cooling, which currently accounts for 30% of peak demand. At the same time, power plant efficiencies (Maulbetsch and DiFilippo 2006) and transmission line (IEEE 2007), substation, and transformer (Askari et al. 2009) capacities are reduced. Higher temperatures increase wildfire activity, which poses risks from direct infrastructure damage to indirect fouling due to soot and firefighting products (Sathaye et al. 2011). These effects are further intensified by loss of hydropower and thermoelectric generation supply (NETL 2010) due to limited water supply.

The Gulf states are affected by a different set and intensity of extreme events, though drought is clearly a threat. Hurricanes are the most apparent and direct hazard for the Southeast and have a wide-ranging impact on the EWL interface as well as society. Hurricanes are a major cause of direct damage, such as destruction of crops, and represent exceptional risks for widespread flooding. In terms of energy, flooding can affect electricity generation and destroy dedicated energy

Sea level rise is another significant hazard for the Gulf states. Sea level rise is characterized in several ways, some through climate and some not. Climate related examples of sea level rise include thermal expansion of the ocean in response to increased atmospheric temperatures. Storm surge and wave action resulting from extreme weather events (e.g., hurricanes) can temporarily induce sea level rise and destruction of property. Other non-climatic events include subduction of land masses (as is occurring in Florida) and storm surge resulting from Tsunami events as was experienced in the spring of 2011 along the west coast of California after the Japanese Tohoku earthquake and tsunami. Coupled with large storms, sea level rise has the potential to significantly inundate large areas of land. In addition to infrastructure impacts, large floods can harm coastal farming land and even induce saltwater intrusion into freshwater aquifers and coastal forests (Karl et al. 2009).

3.3.5 Institutions and Adaptive Capacity

Each region's unique characteristics and institutions dictate the relative capacities of candidate solutions to mitigate and/or adapt to climate change (Figures 3.5-3.6). For example, solar energy may be a good mitigation choice for some regions while being a poor choice in others. How adaptation strategies and associated institutions evolve has significant implications for energy-water-land dynamics.

California is recognized as a national leader in planning toward energy efficiency, environmental stewardship, and water resource management. This aggressive planning is driven by a mix of necessity, limited resources coupled with concentrated growth, and environmental consciousness. Specifically, the state has taken the lead in recognizing and planning for climate change. The Global Warming Solutions Act of 2006, Assembly Bill 32, (Núñez, Chapter 488, Statutes of 2006) sets an economy-wide cap on California GHG emissions at 1990 levels by no later than 2020. This is an aggressive goal that represents approximately an 11% reduction from current emissions levels and nearly a 30% reduction from projected business-as-usual levels in 2020. The second important piece of climate change legislation from 2006 is Senate Bill 1368 (Perata, Chapter 598, Statutes of 2006), which requires the Public Utilities Commission and the Energy Commission to implement an emissions performance standard for all retail providers of electricity in the state. Additionally, California is leading the development of the Western Climate Initiative (WCI), which is designing a "cap and trade" system for the West.

Toward achieving emission standards for retail providers of electricity, California has set aggressive Renewable Portfolio Standards. Established in 2002 under Senate Bill 1078, accelerated in 2006 under Senate Bill 107 and expanded in 2011 under Senate Bill 2, California's Renewables Portfolio Standard (RPS) is one of the nation's most ambitious renewable energy standards. The RPS program requires investor-owned utilities, electric service providers, and community choice aggregators to increase procurement from eligible renewable energy resources to 33% of total procurement by 2020. Unlike the Southeast, California has a wide range of options to meet the RPS. Specifically, the state is pursuing a mix of roughly 30% wind, 35% solar, 20% geothermal, 10% biomass, and 5% small hydroelectric (California Energy Commission 2011). Much of the rest of new electric capacity will be met with natural gas. Except in the case of thermal solar and geothermal generation, the selected mix of generation promotes relatively efficient use of water in electricity generation (Macknick et al. 2011). This, coupled with the fact that the state is aggressively encouraging use of alternative water sources for thermoelectric development (see above) should help alleviate competition across the energy-water and land-water (e.g., municipal and agricultural water use) bilateral interfaces. However, choices favoring wind, solar, geothermal, and natural gas (when considering energy extraction) mean a larger land footprint. This is particularly important for solar, where significant development in sensitive desert environments is likely. Finally, it is important to note that these water and land implications are not limited to California, as roughly 34% of its electricity and 85% of its natural gas supplies are produced outside the state (California Energy Commission 2008).

In terms of reducing CO_2 emissions to the atmosphere, the Gulf states have a different set of options than California. For example, the Southeast has low potential for solar and wind power due to poor insolation and generally low winds; this can be seen in terms of almost nonexistent large wind and solar installations (EPA 2011a). Geothermal potential is largely absent too. However, a proliferation of deep saline aquifers (for sequestering CO_2) and coal-fired power makes the Gulf states a strong candidate for carbon capture and storage (see Illustration 3.2). As an example, Middleton et al. (2012b) identified a set of seven possible geologic reservoirs capable of handling more than 1000 $MtCO_2$/yr from 20 coal-fired power plants across four Gulf states. Retrofitting an existing pulverized coal power plant for CCS approximately doubles water consumption per megawatt-hour generated while reducing generation efficiency by around one-third. Consequently, coal-fired power plants with CO_2 capture will consume more water to produce the same amount of electricity or be required to purchase makeup electricity, or both. Storing vast quantities of CO_2 in the subsurface will also affect the EWL interface by requiring large tracts of the subsurface as well as land for surface access. For example, 50 years of storing 1 $GtCO_2$/yr under typical geologic conditions (e.g., CO_2 density of 37.5 lb/ft^3, 165 ft thick formation, 15% porosity, and an efficiency factor of 0.05) would require an area of 87,000 mi^2 of surface/subsurface land across the continental U.S., equivalent to the area of Minnesota.

Illustration 3.2: Climate-Related Decision Making at the Nexus
Carbon Capture and Storage and Coal-Fired Power Plants

Over 50% of electricity in the U.S., and 30% of greenhouse gas emissions, come from coal-fired power plants (EIA 2011; EPA 2011a). The addition of carbon capture and storage (CCS) to these facilities is a climate change mitigation technology potentially capable of reducing atmospheric greenhouse emissions. Currently, there are 30 small-scale CCS facilities operating in the U.S. as of April, 2011.

CCS captures and compresses CO_2 at large industrial sources (e.g., coal-fired power plants, biorefineries, cement works), transports the CO_2 in a dedicated pipeline network, and injects it in a geological reservoir for storage (e.g., depleted oil and gas fields, deep saline aquifers).

For CCS to have a meaningful impact on GHG emissions, the U.S. will have to capture, transport, and store billions of tonnes of CO_2 in the coming decades (Middleton et al. 2012a). This translates to capturing CO_2 from hundreds of power plants and constructing a pipeline network capable of handling as much as 25% more CO_2 by volume than present day oil consumption (Middleton et al. 2012a).

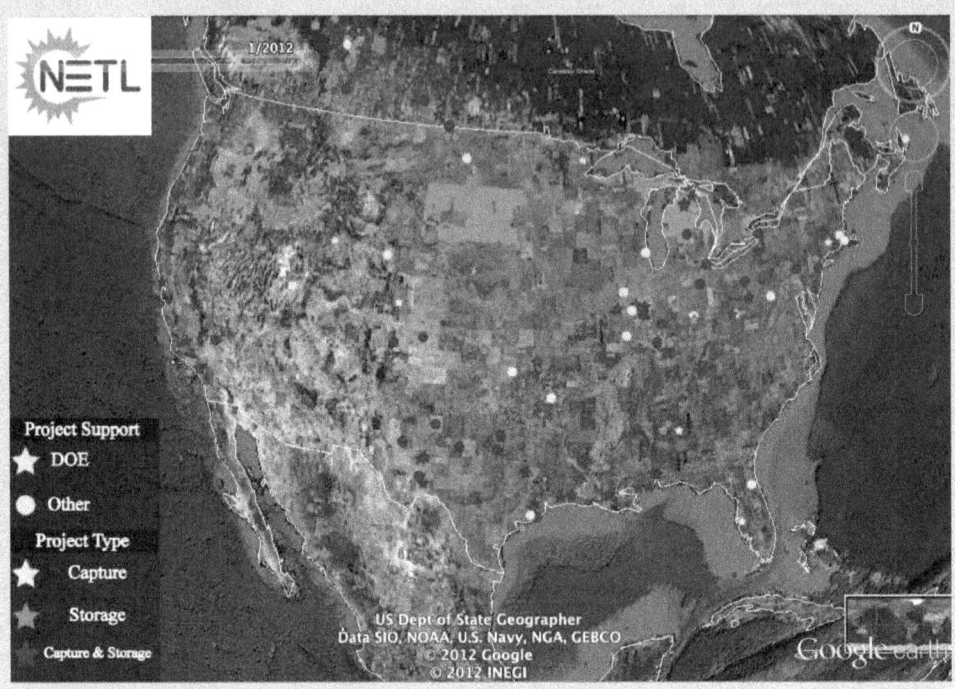

Figure 3.7. Locations of active, proposed, cancelled, and terminated carbon capture and storage projects in the United States as of April 2011. Data from the NETL (2012).

Consideration in Decision Making:

Integrating CCS into the U.S. electricity and carbon portfolio involves several tradeoffs. Coal-fired power plants that use evaporative cooling and are fitted with CCS would consume twice as much water per unit of electricity generated as a non-CCS coal-fired facility (Zhai et al. 2011). Existing water requirements for thermoelectric cooling are already problematic in some regions of the U.S. (See Climate-Related Decision Making at the Nexus, Illustration 2.1). The long-term availability of water for CCS facilities is uncertain, as the increasing demands from a growing population collide with long-term changes in supplies driven by climate change.

Indeed, dry cooling is an option. For a plant using air-cooled condensers, water use for plants without CCS would be reduced by about 80%, while plants with CCS would be reduced by only about 40% – but the capital cost would approximately triple (Zhai et al. 2011). This is significant given that the addition of CCS to operating coal-fired facilities would account for approximately 80% of operating costs; and 20–30% of the costs for new, integrated gasification combined cycle (IGCC) facilities (DOE 2012).

Approximately 40% of coal-fired power plants are located over geologic formations appropriate for CCS (DOE 2012), and retrofitting CCS would have a relatively small land use footprint compared with construction of new facilities. However, the electricity generation efficiency penalty for CCS could increase the requirements for coal by 25–40% (IPCC 2005). This would require more extraction, and concomitant impacts on health, water (from mine tailings and drainage), and land use.

Unlike the Southeast, California's biofuel potential is relatively limited (Figure 3.8), largely because cultivation of energy crops would require irrigation (Department of Energy 2011). One opportunity for California is in the production of forest biomass (integrated forest operations, thinnings, and mill process residues). Use of these feedstocks has multiple implications across the EWL interfaces. First, significant lands would be managed in part for energy production, potentially fueling rural development. The California state water plan recognizes that forest thinning has multiple land-water implications, including improving the water supply reliability, protecting water quality, increasing flood protection, and promoting environmental stewardship (California Department of Water Resources 2009). Beyond the land-water interface, better supply reliability will likewise improve reliability of hydroelectric production. Ultimately, that fraction of energy produced through forest wastes means less land, water, and primary fuel use through other, more resource-intensive electricity generation methods.

The southeast Gulf states generally produce fewer agricultural products than California. For example, all California agriculture commodities generated receipts of $37.5 billion in 2010, whereas the Gulf states combined produced $27.4 billion (USDA 2012). This is approximately 27% less (by value), even though the Gulf states have a combined area 1.7 times larger than California. However, the Gulf states have a much greater potential for biofuels (Figure 3.8). For example, NREL (2005) estimates that California has potential to produce 13.4 million tonnes (13.4 MT) of biomass (includes dedicated energy crops, agricultural residues, municipal discards, and wood residues); by comparison, the Gulf states could produce 52.5 MT. Due to the availability of rain-fed agriculture, biomass and biofuel production in the Southeast will have limited impact on water availability though water quality will certainly be significant.

Electricity generation in California may become even more important due to Assembly Bill 1493, which requires a 30% reduction in GHG emissions from vehicles sold in California by 2016. One the most promising options for reaching this goal involves increased use of plug-in hybrid electric vehicles and all-electric vehicles (California Energy Commission 2011). Also realized is the need to reduce the number of vehicle miles traveled in the state, potentially by locating homes closer to workplaces, and increasing public transportation options and use.

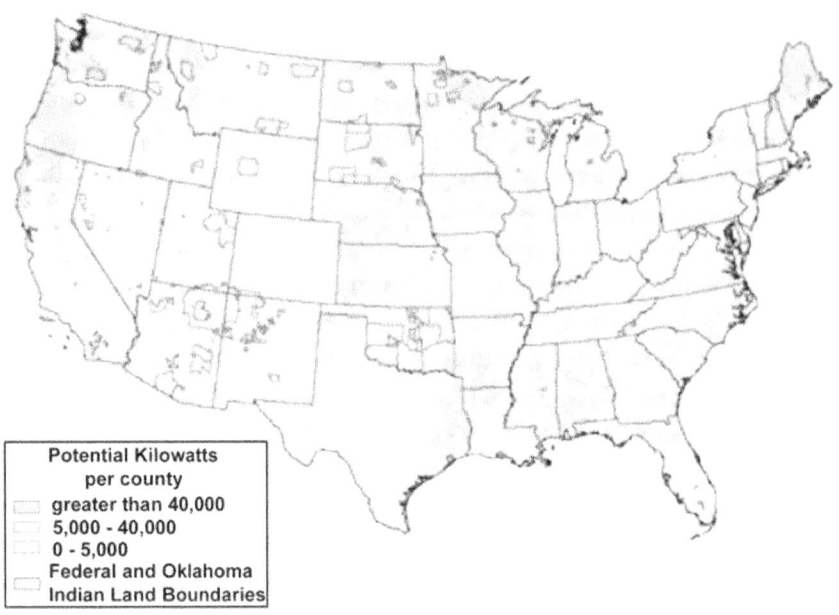

Potential Kilowatts
per county
☐ greater than 40,000
☐ 5,000 - 40,000
☐ 0 - 5,000
☐ Federal and Oklahoma
Indian Land Boundaries

Figure 3.8. Biomass and biofuels resource potential in the conterminous United States (National Atlas, 2012).

California is also aggressively pursuing resource conservation. As an example, per capita electricity use has remained relatively stable over the past 30 years while nationwide use has increased by about 50% (California Energy Commission 2008). Efficiency gains have been made principally through building codes, appliance standards, and utility standards. Reduced electricity use mean reductions in future water demands and land brought under energy development (both for power plants and energy extraction). Improved water efficiencies are hoped to yield reductions of up to 1 MAF in the agricultural sector and 3.1 MAF in the urban sector (California Department of Water Resources 2009). Such savings are significant in terms of energy used to capture, convey, treat, and distribute the water. Specifically, California consumes 30% of the electricity and 20% of the natural gas for these functions (including end use applications). However, the state also has plans to augment water supply through desalination (0.4 MAF) and recycled municipal wastewater (2.3 MAF), which represent new energy demands on the system (California Department of Water Resources 2009).

3.3.6   Summary of Regional Differences

California and the Gulf states are characterized by very different physiographies, economies and availability of natural resources; consequently, each region will be differentially impacted by climate and each region will have to adapt/mitigate using different responses. Differences in these responses have important implications for the evolving dynamics between the EWL interfaces (compare Figures 3.5 and 3.6).

California and the Gulf states would likely follow very different compliance paths if stringent emission standards were adopted. Because of abundant renewable resources and past proactive adoption on the part of the state, California would expand electricity production with wind, solar, geothermal, biomass, and small hydroelectric. In contrast, limited renewables in the Gulf states my lead to broad implementation of CCS utilizing their abundant deep saline aquifers for storage. These different strategies also reflect differences in water resources between the two regions. Utilization of renewables coupled with the state's restriction on using potable water for new electricity generation should yield minimal impact on California's limited water resources. On the other hand, utilization of CCS will require expanded water withdrawals in the Gulf states, which the region is much more able to accommodate (such new demands could be minimized through conversation of open-loop cooling to closed-loop). The nexus of water and renewable resources also plays an important role in the choice of transportation fuels. Because of limited water resources and arable land, California would likely turn to plug-in hybrid electric vehicles and all-electric vehicles. In contrast, the availability of water and land in the Gulf states would favor cultivation of biomass for biofuel production. This is not to say that California would not utilize biomass; rather, the state is likely to develop forest feedstocks for electricity generation, which in turn can be used to reduce wildfire danger and improve watershed production. The combined action on the part of California would help manage the strong competition over water, which is less of an issue for the Gulf states.

3.4 References

Ana RS. 2011. "South Texas Farmers Bracing for Water Shortages." *AgriLife Today*, December 12, 2011. Accessed at: http://agrilife.org/today/2011/12/28/south-texas-water-shortages/.

Askari MT, A Kadir, WFW Ahmad, and M Izadi. 2009. "Investigate the Effect of Variations of Ambient Temperature on HST of Transformer." *Proceedings of the 2009 IEEE Student Conference on Research and Development*, November 16-18, 2009, pp. 363-367, DOI: 0.1109/SCORED.2009.5442998.

Averyt, K, J Fisher, A Huber-Lee, A Lewis, J Macknick, N Madden, J Rogers, and S Tellinghuisen. 2011. Freshwater Use by U.S. Power Plants: Electricity's Thirst for a Precious Resource, A Report of the Energy and Water in a Warming World Initiative. Union of Concerned Scientists, Cambridge, Massachussetts.

AWEA – American Wind Energy Association. 2012. 4th Quarter 2011 Public Market Report. Washington, DC.

California Department of Water Resources. 2009. California Water Plan Update 2009, Bulletin 160-09. Accessed at: http://www.waterplan.water.ca.gov/cwpu2009/index.cfm.

California Energy Commission. 2011. RPS Program Update. Accessed at: http://www.cpuc.ca.gov/PUC/energy/Renewables/hot/RPS+Program+Update.htm.

California Energy Commission. 2008. Energy Action Plan: 2008 Update. February 2008. Accessed at: http://www.energy.ca.gov/energy_action_plan/.

California Public Utilities Commission. 2011. Embedded Energy in Water Studies 1, 2 and 3, May 10, 2011. Accessed at: http://www.cpuc.ca.gov/PUC/energy/Energy+Efficiency/EM+and+V/Embedded+Energy+in+Water+Studies1_and_2.htm.

Cayan DR, MD Dettinger, RT Redmond, GJ McCabe, N Knowles, and DH Peterson. 2003. "The Transboundary Setting of California's Water and Hydropower Systems, Linkages Between the Sierra Nevada, Columbia, and Colorado Hydroclimates." Chapter 11 in *Climate and Water: Transboundary*

Challenges in the Americas, eds. H Diaz and BJ Morehouse. Kluwer Academic Publishers, Dordrecht/Boston.

Christensen N and DP Lettenmaier. 2006. "A Multimodel Ensemble Approach to Assessment of Climate Change Impacts on the Hydrology and Water Resources of the Colorado River Basin." *Hydrology and Earth System Sciences* 3(6):3727-3770.

Dawson B. 2012. "A Drought for the Centuries: It Hasn't Been This Dry in Texas Since 1789." *Texas Climate News*, February 8, 2012. Accessed at: http://texasclimatenews.org/wp/?p=3355.

DOE – U.S. Department of Energy. 2011. U.S. billion-ton update: Biomass supply for a bioenergy and bioproduction industry. Prepared by Oakridge National Laboratory, August 2011, 194 pp.

DOE – U.S. Department of Energy. 2009. Modern Shale Gas Development in the United States: A Primer, April 2009. Accessed at: http://www.gwpc.org/e-library/documents/general/Shale%20Gas%20Primer%202009.pdf

DOE – U.S. Department of Energy. 2012. Post-Combustion Capture Research. Accessed at: http://www.fossil.energy.gov/programs/powersystems/pollutioncontrols/Retrofitting_Existing_Plants.html

Energy Collective. 2011. Brown's Ferry Nuclear Power Plant Has to Shut Down Again. Accessed at: http://theenergycollective.com/jcwinnie/62883/brown-s-ferry-nuclear-power-plant-has-shut-down-again.

EPA - U.S. Environmental Protection Agency. 2011a. Emissions & Generation Resource Integrated Database (eGRID). Accessed at: http://www.epa.gov/cleanenergy/energy-resources/egrid/index.html.

EPA - U.S. Environmental Protection Agency. 2011b. Investigation of Ground Water Contamination Near Pavillion Wyoming. Draft Report, December 2011. Accessed at: http://www.epa.gov/region8/superfund/wy/pavillion/EPA_ReportOnPavillion_Dec-8-2011.pdf.

ERCOT - Electric Reliability Council of Texas. 2011. Grid Operations and Planning Report. Accessed at: http://www.ercot.com/content/meetings/board/keydocs/2011/1212/Item_06e_-_Grid_Operations_and_Planning_Report.pdf.

EIA – U.S. Energy Information Administration. 2011. Annual Energy Outlook 2011 with Projections to 2035. Accessed at: http://www.eia.gov/oiaf/aeo/gas.html

EIA – U.S. Energy Information Administration. 2008. Summary Maps: Natural Gas in the Lower 48 States and North America. Accessed at: http://205.254.135.7/pub/oil_gas/natural_gas/analysis_publications/maps/maps.htm.

Fannin B. 2011. "Texas agricultural drought losses reach record $5.2 billion." *AgriLife TODAY*. August 17, 2011. Accessed at: http://agrilife.org/today/2011/08/17/texas-agricultural-drought-losses-reach-record-5-2-billion/.

Fernandez M. 2012. "Texas Drought Forces a Town to Sip From a Truck." *The New York Times*, February 4, 2012. Accessed at: http://www.nytimes.com/2012/02/04/us/texas-drought-forces-town-to-haul-in-water-by-truck.html?_r=2.

Georgakakos AP, NE Graham, F -Y Cheng, C Spencer, E Shamir, KP Georgakakos, H Yao, and M Kistenmacher, 2012. "Value of Adaptive Water Resources Management in Northern California Under Climatic Variability and Change." *Journal of Hydrology* 412-413(2012): 34-46, DOI: 10.1016/j.jhydrol.2011.04.038.

Harou JJ, J Medellín-Azuara, T Zhu, SK Tanaka, JR Lund, S Stine, MA Olivares, and MW Jenkins. 2010. "Economic Consequences of Optimized Water Management for a Prolonged, Severe Drought in California." *Water Resources Research* 46:1-12, DOI: 10.1029/2008WR007681.

IEEE – Institute of Electrical and Electronics Engineering. 2007. Standard for Calculating the Current-Temperature of Bare Overhead Conductors. IEEE 738-2006, Institute of Electrical and Electronics Engineers, New York.

IPCC. 2007. Climate Change 2007: Synthesis Report. Contribution of Working Groups I, II and III to the Fourth Assessment Report of the Intergovernmental Panel on Climate Change [Core Writing Team, RK Pachauri and A Reisinger]. IPCC, Geneva, Switzerland, 104 pp.

IPCC. 2005. Bert Metz, Ogunlade Davidson, Heleen de Coninck, Manuela Loos and Leo Meyer (Eds.) Cambridge University Press, UK. pp 431.

Karl TR, J M Melillo, and TC Peterson. 2009. Global Climate Change Impacts in the United States, eds. TR Karl, JM Melillo, TC Peterson. Cambridge University Press, Cambridge, England.

Kenny JF, NL Barber, SS Hutson, KS Linsey, JK Lovelace, and MA Maupin. 2009. Estimated Use of Water in the United States in 2005, U.S. Geological Survey Circular 1344. U.S. Geological Survey, Reston, Virginia.

Macknick J, R Newmark, G Heath, and KC Hallett. 2011. A Review of Operational Water Consumption and Withdrawal Factors for Electricity Generating Technologies, NREL/TP-6A20-50900. National Renewable Energy Laboratory, Golden, Colorado.

Malewitz J. 2011. "Persistent drought threatens Texas oil industry." *Stateline*, October 7, 2011. Accessed at: http://www.stateline.org/live/details/story?contentId=605095

Maulbetsch JS and MN DiFilippo. 2006. *Cost and Value of Water Use at Combined-Cycle Power Plants.* CEC-500-2006-034. California Energy Commission.

Middleton RS, GN Keating, PH Stauffer, HS Viswanathan, and RJ Pawar. 2012a. "Effects of geologic reservoir uncertainty on CCS infrastructure." *International Journal of Greenhouse gas Control*, in review.

Middleton RS, GN Keating, PH Stauffer, A Jordan, HS Viswanathan, Q Kang, JW Carey, M Mulkey, EJ Sullivan, S Chu, and RA Esposito. 2012b. "The multi-scale science of CO_2 capture and storage: From pore scale to regional scale." *Energy and Environmental Science*, in review.

Milly PCD, J Betancourt, M Falkenmark, RM Hirsch, ZW Kundzewicz, DP Lettenmaier and R J.Stouffer. 2008. "Stationarity is dead: Whither water management?" *Science* 319:573-574.

National Atlas. 2012. Accessed at: http://www.nationalatlas.gov/.

National Climate Assessment Report. 2013a. Climate Change and Infrastructure, Urban Systems and Vulnerabilities Data Analysis Report.

National Climate Assessment Report. 2013b. Southwest Regional Data Analysis Report.

National Drought Policy Commission. 2004. Texas State Drought Program. Available at: http://govinfo.library.unt.edu/drought/finalreport/filec/Texas%20State%20Drought%20Programs.htm.

National Energy Education Development Project. 2008. "Hydropower," in *Secondary Energy Infobook*, pp. 24-27. NEED Project, Manassas, Virginia.

NETL – National Energy Technology Laboratory. 2010. Water Vulnerabilities for Existing Coal-fired Power Plants. DOE/NETL-2010/1429, National Energy Technology Laboratory, Pittsburgh, Pennsylvania.

NREL – National Renewable Energy Laboratory. 2005. Biomass Maps. Accessed at: http://www.nrel.gov/gis/biomass.html

Newell, R. 2011. Statement of Richard Newell Administrator, EIA U.S. DOE before the Committee on Natural Resources U.S. House of Representatives, Written Testimony to Congress, March 17, 2011.

Nielsen-Gammon JW. 2011. The 2011 Texas Drought: A Briefing Packet for the Texas Legislature. The Office of the State Climatologist, College Station, Texas.

Nyberg M. 2009. 2008 Net System Power Report. CEC-200-2009-010. California Energy Commission

Palmer W. 1965. Meteorological Drought. Research paper no.45, U.S. Department of Commerce Weather Bureau. 58pp. Available online by the NOAA National Climatic Data Center at http://www.ncdc.noaa.gov/temp-and-precip/drought/docs/palmer.pdf

Pacala S and R Socolow. 2004. "Stabilization Wedges: Solving the Climate Problem for the Next 50 Years with Current Technologies." *Science* 305(5686): 968-972, DOI: 10.1126/science.1100103.

Plushnick-Masti R 2011. "Water: West Texas coal plant finds a water supplier." AP/*Houston Chronicle*, July 12, 2011.

Sathaye J, L Dale, P Larsen, G Fitts, K Koy, S Lewis, and A Lucena. 2011. Estimating Risk to California Energy Infrastructure from Projected Climate Change. CEC-500-2011-XXX. California Energy Commission.

Smith R. 2011. "Texas power grid falls short." *Wall Street Journal*, August 12, 2011. Accessed at: http://online.wsj.com/article/SB10001424053111904823804576502592393033486.html.

Stauffer PH, GN Keating, RS Middleton, HS Viswanathan, KA Berchtold, RP Singh, RJ Pawar, and A Mancino. 2011. "Greening Coal: Breakthroughs and Challenges in Carbon Capture and Storage." *Environmental Science and Technology* 45(20)8597–8604, DOI: 10.1021/es200510f.

Stillwell AS, ME Clayton, and ME Webber. 2011. "Technical Analysis of a River Basin-Based Model of Advanced Power Plant Cooling Technologies for Mitigating Water Management Challenges." *Environmental Research Letters* 6(3):1–11, DOI: 10.1088/1748-9326/6/3/034015.

Strzepek K, G Yohe, J Neumann, and B Boehlert. 2010. "Characterizing Changes in Drought Risk for the United States from Climate Change." *Environmental Research Letters* 5(4):1–9, DOI: 10.1088/1748-9326/5/4/044012.

Texas Forest Service. 2011a. Preliminary Estimates Show Hundreds of Millions of Trees Killed by 2011 Drought, December 19, 2011. Accessed at: http://txforestservice.tamu.edu/main/popup.aspx?id=14954.

Texas Forest Service. 2011b. Dangerous Wildfire Conditions Predicted for Friday. Accessed at http://txforestservice.tamu.edu/main/popup.aspx?id=14644.

Texas Water Development Board. 2012. 2012 State Water Plan. Accessed at: http://www.twdb.state.tx.us/wrpi/swp/swp.asp.

UCS – Union of Concerned Scientists. 2007. Rising Temperatures Undermine Nuclear Power's Promise, Washington, DC. Accessed at: http://www.nirs.org/climate/background/ucsrisingtemps82307.pdf.

UCS – Union of Concerned Scientists. 2012. Accessed at: http://www.ucsusa.org/.

USDA – United States Department of Agriculture. 2012. State Fact Sheets. Accessed at: http://www.ers.usda.gov/StateFacts/.

Vicuña S, R Leonardson, MW Hanemann, LL Dale, and J A Dracup. 2008. "Climate Change Impacts on High Elevation Hydropower Generation in California's Sierra Nevada: A Case Study in the Upper American River." *Climatic Change* 87 (Suppl 1): S123–S137.

White E. 2011. Managing in the southern Plains: Cattle impacts. Accessed at: http://www.drought.gov/imageserver/NIDIS/DEWS/southern_plains/docs/Drought_Webinar_Dec_08_2011.pdf (webinar dated December 8, 2011).

Zhai, H, ES Rubin, and PL Versteeg. 2011. "Water Use at Pulverized Coal Power Plants with Postcombustion Carbon Capture and Storage." *Environ. Sci. Technol.* 45(6):2479–2485, DOI: 10.1021/es1034443.

4.0 Risk, Uncertainty and Vulnerability Associated with Climate Impacts on Energy-Water-Land Interfaces

4.1 Introduction and Background

This section describes the unique characteristics of risk, uncertainty, and vulnerability within and across multi-sector systems to give context to the problem and illustrate the deep complexity of this difficult subject. In researching this topic, it was quickly evident that there is a gap in the literature regarding the makeup and character of cross-sector risk, uncertainty, and vulnerability and descriptions of what makes them unique compared to more focused applications (e.g., single sector or direct sector-to-sector dynamics). Thus, the objective here is to create a comprehensive description of multi-sector risk, uncertainty, and vulnerability that can serve as a foundation for future assessments and research and as a template to help policy and decision makers better understand their implications.

The descriptions of uncertainty and risk in this section are modeled after Morgan et al. (2009), Mastrandrea et al. (2010), and Moss and Yohe (2011), who present a common approach and language for evaluating and communicating the degree of confidence in the NCA. Figure 4.1 depicts the confidence language being used in the NCA. In this framework, confidence is a function of the type, amount, quality, and consistency of evidence (rated as strong, moderate, suggestive, or inconclusive) and the degree of consensus among experts in how to interpret the available evidence. The Moss and Yohe (2011) document synthesizes and refines background material previously developed to support the third, fourth, and fifth assessment reports (Moss and Schneider 2000; Manning et al. 2004; Mastrandrea et al. 2011). A more recent discussion on this topic from many of the same authors is presented in a focused issue of *Climatic Change* titled "Special Issue: Guidance for Characterizing and Communicating Uncertainty and Confidence in the Intergovernmental Panel on Climate Change" (Yohe and Oppenheimer 2011). Collectively, these documents thoroughly define uncertainty, vulnerability, and risk, and readers seeking more detail are encouraged to review those documents.

For this report, risk is defined as the product of the consequence of an event and the likelihood that it will occur, with *total risk* being the integration of risk across all events and probabilities. Uncertainty contributes to risk by widening the distribution of likelihood, which creates the general relationship of the higher the uncertainty, the higher the risk (Backus et al. 2010). Low-likelihood, high-consequence events can dominate risk since high-likelihood, low-consequence events are typically mitigated by system safeguards and resiliency, whereas the consequences associated with low-likelihood events typically happen when a key system is stressed beyond its breaking point. Before exceedance of its breaking point, there may be little or no consequence. The events surrounding Hurricane Katrina are a perfect example of this.

Vulnerability is defined as a function of exposure, sensitivity, and the capacity to adapt, and is key to assessing consequence. Vulnerability is influenced by an event's magnitude and timing, as well as the persistence, reversibility, distribution, and likelihood of the consequence. The potential to adapt to the consequence and the importance (real or perceived) of the consequence also contribute to vulnerability. Resilience and vulnerability are closely linked.

Confidence Level	Example combinations of factors that could contribute to this confidence evaluation
High	Strong evidence (established theory, multiple sources, consistent results, well documented and accepted methods, etc.), high consensus
Medium High	Moderate evidence (several sources, some consistency, methods vary and/or documentation limited, etc.), medium consensus
Medium Low	Suggestive evidence (a few sources, limited consistency, models incomplete, methods emerging, etc.), competing schools of thought
Low	Inconclusive evidence (limited sources, extrapolations, inconsistent findings, poor documentation and/or methods not tested, etc.), disagreement or lack of opinions among experts

Figure 4.1. Confidence levels derived from level of evidence and degree of consensus. Reproduced and adapted from Moss and Yohe (2011).

Some potential impacts could be of such high consequence for society that stakeholders consider them to be "key vulnerabilities" because of their magnitude, timing, persistence/irreversibility, limited potential for adaptation, distributional aspects, likelihood, or other attributes. For these key vulnerabilities, the NCA will be estimating the risks presented as clearly as possible. This involves (1) using a well-defined metric to describe the consequences of the impact (quantitatively, if possible) and (2) using standardized terms/ranges to estimate the likelihood the impact will occur. These standardized probability ranges are shown in Figure 4.2.

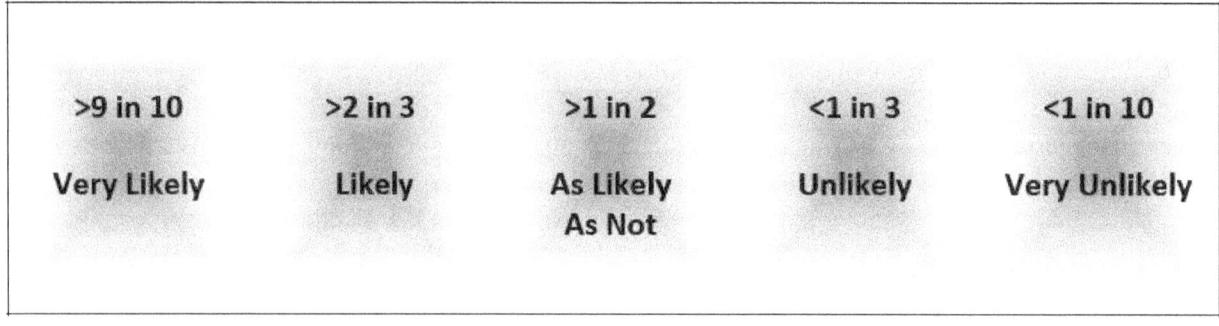

Figure 4.2. Standardized likelihood expressions. These expressions are recommended for use in the NCA to describe how likely it is that a particular consequence will materialize under climate change associated with the B1 and A2 scenario. The technical guidelines document suggests using both the numerical ranges and the likelihood terms together, but not using the terms alone since they have no standardized meaning without the quantitative calibration.

4.2 Risk, Uncertainty, and Vulnerability at the Sector Interfaces

For sector interfaces, the need to address risk, uncertainty, and vulnerability is directly related to the DET model presented in section 2.1.3. Within a single sector, stresses on a system can change demand, endowments (e.g., water used for energy), or both, with little or no consideration of technological transitions or deployments. In this context, technology is defined as the physical conversion of an endowment or the economic production structures that define how an endowment is used by another sector (e.g., agriculture commodity markets). However, for bilateral or across all interfaces of the energy-water-land system, technology has its own set of uncertainties, vulnerabilities, and risks. The challenge is that both likelihood and consequence are more difficult to estimate, as are the associated vulnerabilities and risks.

Unfortunately, risk, uncertainty, and vulnerability at the interfaces are minimally represented in the literature, and where they are represented they are usually specific to a particular inter-sectorial dynamic (e.g., Koutsoyiannis et al. 2009; Falloon and Betts 2010; Miles et al. 2010; Hunt and Watkiss 2011). Collectively, however, the literature is deep enough to describe the general character of risk, uncertainty, and vulnerability at the interfaces. From this, four characteristics have been identified:

- They are broader in scope.

- They can be amplified or attenuated across sectors.

- They have variable and likely non-linear temporal and spatial dynamics.

- They manifest during low-likelihood, high-consequence events.

Understanding these four characteristics is fundamental to understanding how risk, uncertainty, and vulnerability relate to each characteristic across sectors, and for developing solutions and strategies within the inter-sector technologies that may reduce their impact or influence.

The balance of this section addresses the four characteristics in more detail, using examples from the literature where possible to illustrate the concepts and support the conclusions. It ends with a summary that identifies key considerations for including risk, uncertainty, and vulnerability in cross-sector decision making and policy formation.

4.2.1 Characteristic 1 – Broader in Scope

To explain this characteristic, it helps to understand the reason why we simulate the effects of climate change, which is to make decisions and formulate policy that is strategically robust, economically sound, and maximally effective. For an analysis to be useful, it must help policy and decision makers answer two simple questions: (1) How disruptive is the climate change; and (2) How disruptive are the solutions? (Socolow 2011). When applied to cross-sector systems analyses, these questions point to two key elements that broaden the scope of risk, uncertainty, and vulnerability as compared to their intra-sector cousins. The first concerns our ability to adequately predict the effect of climate change and its associated risks, uncertainties, and vulnerabilities (van Pelt and Swart 2011). The second concerns feedback across the interfaces, which can cause secondary (but not necessarily smaller) consequences that ripple across the various social, economic, and natural environments.

Falloon and Betts (2010) provide a real world example of these elements as applied to the water and agriculture (land) sectors, by identifying a set of uncertainties about the effects of climate on water and agriculture that includes the sector-to-sector interactions themselves. This implies that risk, uncertainty, and vulnerability within multi-sector systems are not limited to a simple propagation from one sector to the next, but include the sector-to-sector dynamics. Freibauer et al. (2004) support this idea by showing how changing agricultural land use to mitigate climate impact is likely to have secondary effects that may be detrimental to the original objective.

Inter-sectorial risk, uncertainty, and vulnerability are not just a function of epistemic (incomplete knowledge) and stochastic (unknowable factors) uncertainties, but also of uncertainties associated with human response (van Pelt and Swart 2011; Backus et al. 2012). Human response is integral to both how we cope with the changing climate and how we respond to mitigation and adaptation policies (van Pelt and Swart 2011). Furthermore, risk, uncertainty, and vulnerability are not necessarily related from sector to sector in the same way that the corresponding demands, endowments, and technologies are (Webber 2011), nor are those relationships temporally constant. The difficulty stems from the additional uncertainty caused by our inability to predict changing feedback dynamics and operational norms of newly implemented policies and regulations (Scott and Pasqualetti 2010; Newell et al. 2011; Scott et al. 2011; Webber 2011). This is addressed further, along with several examples, in the discussion of Characteristic 2 in section 4.2.2.

When applied to cross-sector systems, the unique inter-connectedness among the three types of uncertainties (epistemic, stochastic, and human response) underlies the constant struggle to place risk, uncertainty, and vulnerability into a decision-making context. The Texas drought case study presented in section 3.2 is a real world example of these concepts. Uncertainty about the drought's duration and magnitude created real and perceived risks for each sector. Because competition for water was the driving dynamic influencing these risks, policy makers began to implement different water and energy conservation policies in hopes of reducing local consequences (e.g., municipal water supplies running low) and regional consequences (e.g., lake levels dropping below power plant intakes) and their associated risks. The near-term sector-to-sector dynamics are mostly steady-state (i.e., unchanging), but over the long term the dynamics can be expected to change. Lasting effects on changes in land use and vegetation will permanently change the dynamics of the hydrologic system. Human attitudes and behavior regarding water use in Texas will change as will their approach for long-term water management. The type and magnitude of the changes are sector specific, with some sectors going back to business as usual and others making significant changes. This illustrates a process whereby risk-based decisions are made using predictions of uncertain future conditions, and where uncertainty about the effectiveness of the decisions adds to the uncertainty in those predictions, creating a feedback loop that is difficult to assess.

It is important to note that the decisions themselves and their effectiveness are functions of human behavior (among other things), which in turn is a function of the risk, both real *and perceived*. Thus, the Texas drought demonstrates how the relationships among the stochastic (e.g., the duration and magnitude of the drought), epistemic (e.g., what are the future energy and water requirements), and human response uncertainties (e.g., to what degree are the water and energy conservation policies adopted) are cumulative functions of one another that make it more difficult to answer questions about the effects of climate change and the effects of the solutions. In other words, risk, uncertainty, and vulnerability across sector interfaces are broader in scope than single sector estimates of the same.

4.2.2 Characteristic 2 – Amplification and Dampening Across Sectors

For this discussion, amplification and attenuation (AA) refers to how risk, uncertainty, and/or vulnerability for systems in one sector react to changes from systems in a different sector. While similar to the definition of sensitivity with regards to numerical modeling, AA across sectors is more comprehensive since the sensitivities are conditional and the dynamics that apply today may not apply tomorrow (Barabasi 2005). This is mainly because the relationships between the uncertainty in one sector and the risk and vulnerabilities in other sectors are highly non-linear and difficult to predict. Here, we describe some key features of this characteristic and the implications for predicting these dynamics.

Dams and reservoirs provide a simple example of AA. During high runoff, dams attenuate peak flow to mitigate the risks of downstream flooding, while during summer low flows, dams may release water to meet downstream environmental or water supply needs. This type of cross-sector AA of a physical endowment such as water represents a primary dynamic between sectors. Primary dynamics refer to the direct reaction of a dependent system in one sector to changes in the resource/endowment supply from another sector. Examples of primary dynamics would include the direct reduction in crop or energy production due to a given water deficit, or the reduction in water use as a result of a water conservation policy.

As discussed further below, our understanding of primary dynamics and our ability to cope with their (typical) variability is relatively good. However, with respect to risk, uncertainty, and vulnerability at the sector interfaces, it is the secondary dynamics that are of most interest. Secondary dynamics are the ripple effects and clustering of activity that result from the primary dynamic(s). Secondary dynamics are executed and felt within the receiving sector and beyond. For instance, a water deficit creates a relatively predictable reduction in crop production (the primary dynamic), which can cause revenue losses for farmers, higher unemployment and less purchasing, and land-use changes (the secondary dynamics). Each secondary dynamic, in turn, produces its own higher order ripple effects.

In a relevant study of AA that focuses on environmental variability and population dynamics, Greenman and Benton (2005) note that variability in environmental stressors can excite otherwise stable population dynamics into large amplitude fluctuations that are sometimes orders of magnitudes larger than the original variability. This concept is identified in similar studies (Blarer and Doebeli 1999; Greenman and Benton 2003; Greenman and Benton 2005). Greenman and Benton (2005) provide a mathematical basis for AA that shows that the dynamics between the stressor(s) and the population(s) can change, as can the population dynamics themselves. We can link this concept to risk, uncertainty, and vulnerability across sectors by assuming that variabilities in the environmental stress from Greenman and Benton (2005) are similar to uncertainties in future climatic conditions, which suggests that uncertainty will influence AA as well as the sector-to-sector technologies (i.e., the dynamics between the sectors). This implies that uncertainty is also amplified or attenuated, which in turn directly influences risk and vulnerability.

It has been argued that human response to primary or secondary dynamics is the main influence on AA across sector interfaces (Renn 2011). This is supported by studies of human response to information flow (Jo et al. 2012), market stresses (Ghoulmie et al. 2005), and network communications (Eckmann, Moses et al. 2004; Barabasi 2005). As a result, the concept of AA is directly addressed by the social sciences using a theory called the Social Amplification of Risk Framework (SARF) (Kasperson et al. 1988). The main thesis of SARF states that impact events interact with individual psychological, social, and other cultural factors to produce AA in the public perceptions of risk. Individual and group responses then generate secondary social and economic impacts while simultaneously amplifying or attenuating the physical risk itself (Kasperson et al. 1988). The SARF includes a description of ripple effects (i.e., secondary dynamics) that are caused by the AA of risk that includes enduring mental perceptions and behavior, economic impacts, changes in property values, training, education, and social disorder. In turn, individuals and groups perceive and react to the secondary dynamics, resulting in higher order dynamics that may ripple to other sectors, populations, and locations. In later work, Kasperson (2006) goes on to argue that traditional risk analyses neglect these secondary and higher order dynamics and thus vastly underestimate the risk of certain impact events.

A simple example of how human perceptions can amplify or attenuate physical risk is found in a farm-scale study that examines farmer responses to the probability that climate change could cause a catastrophic loss to their crops (Alpizar et al. 2011). The study showed that the farmers were more willing to adopt new tactics or invest in new strategies if the probability of catastrophic loss *was unknown or uncertain than if the probability was known*. For example, the farmers were more willing to adapt to climate change when the probability was communicated as a range (e.g., 5–30%) than as a single value (e.g., 30%). Mathematically, both statements are virtually identical since a 30% probability of loss includes the chance of loss for probabilities less than 30%. In this case, the risk was amplified by the farmers' perceptions and responses to a single value of probability by the fact that they were less likely to take precautions, making them more vulnerable to loss and increased risk.

A final example that highlights how changes in sector-to-sector dynamics can amplify risk, uncertainty, and vulnerability can be found in the real estate market crash of 2008. A significant contributor to the market crash was a set of policies and regulations implemented in the late 1990's and early 2000's that created a new operating landscape by introducing new players while also reducing the checks and balances that controlled the old operating landscape (Martin 2011). The consequences of the crash reached far beyond the real estate market to impact nearly every sector of the global economy (Martin 2011). Known as "market liberalization," this sudden change in the operating landscape can amplify multi-sector dynamics beyond our ability to predict them and drive the relevant system(s) to new (and unknown) operational norms (Weil 2010). While the term "market liberalization" is likely not appropriate when applied to climate change, its meaning is. Within climate change, changes in the operating landscape can occur from changes in the physical environment, the introduction of new policies and regulations, or both, that can create new trigger points where the system dynamics suddenly change. The consequence is that changes in the operational landscape represent another level of risk through their potential to amplify uncertainty and vulnerability for systems and sectors far beyond the intended scope. This concept relates to the question presented above about the impacts of the solutions. Here we see where the impacts were amplified and extended in unpredictable and unintuitive ways.

4.2.3   Characteristic 3 – Increased Temporal and Spatial Dynamics

Because the intersection between climate, energy, water, and land exists on a global scale, the events in one part of the world can have large impacts elsewhere. This is a result of the inter-sectorial technologies and the integrating effect that the atmosphere and climate imposes on the entire world. When estimating risk within a single sector, the effects and consequences are generally limited to adjacent mechanisms in time and space. For instance, the effect of a drought when analyzed from a single sector point of view would include the reduction in production (e.g., agricultural, energy) for the direct users of that endowment. However, when expanded to multi-sector systems, the temporal and spatial scales can increase. To use the drought example, the immediate reduction in agricultural production could impact food prices in another part of the world. Furthermore, those impacts could be delayed due to existing inventories or changes in demand.

Hoff (2009) presents several examples of these larger temporal and spatial dynamics that result from the worldwide effect of rapid deforestation in the Amazon, the competition between biofuel feedstock and food production, and regional to international scale institutional policies and regulations. These larger scale dynamics present yet another layer of risk, uncertainty, and vulnerability by forcing any analysis to consider how these scales influence and change the inter-sectorial dynamics. For instance, in water resource management, the effects of climate change at a local scale (sub-watershed) may be quite different than those at a regional scale (basin to multi-basin) as could be the inter-sector linkages that convert these effects to consequences in other sectors. Furthermore, the management structure and solution space can also vary with scale, such as a national water policy administered by the federal government that influences local water management.

Another aspect of scale is the spatial interdependencies that develop in one sector due to changes in another sector. Backus et al. (2010) looked closely at this spatial component by assessing the near-term risk of climate uncertainty and the interdependence among the U.S. states. The study uses the range of predicted precipitation over the contiguous U.S. (CONUS) from 53 climate model simulations (from 24 different models) from the IPCC AR4 to represent uncertainty in climate predictions for the 40 years from 2010 to 2050. Using an integrated energy-water-agriculture model (Tidwell et al. 2009), the range of precipitation predictions are used to simulate competition for water and the resulting water scarcity for 7 different sectors: thermo-electric power, hydro-electric power, municipal, industrial, mining, agriculture (limited to corn and soy), and livestock. Water scarcity for each sector is converted into costs for 73 economic sectors using the Regional Economic Models Incorporated (REMI) macroeconomic model (REMI 2009). The results are analyzed to determine the total risk to the gross domestic product (GDP), employment rate, and population of each state. Their results show 42 states with minimal to significant (>-0.50%) negative impacts to their GDP risk and 6 states with a net positive change to their GDP risk (Figure 4.3). Most of the variability among the states is attributed to how consumers and industries respond to the changing conditions rather than the localized effects of climate change. In addition, the results show that the *relative* effects are important in that jobs and the population tend to move to states that have better economic conditions than their present state, even if the economic condition in the receiving state is also down.

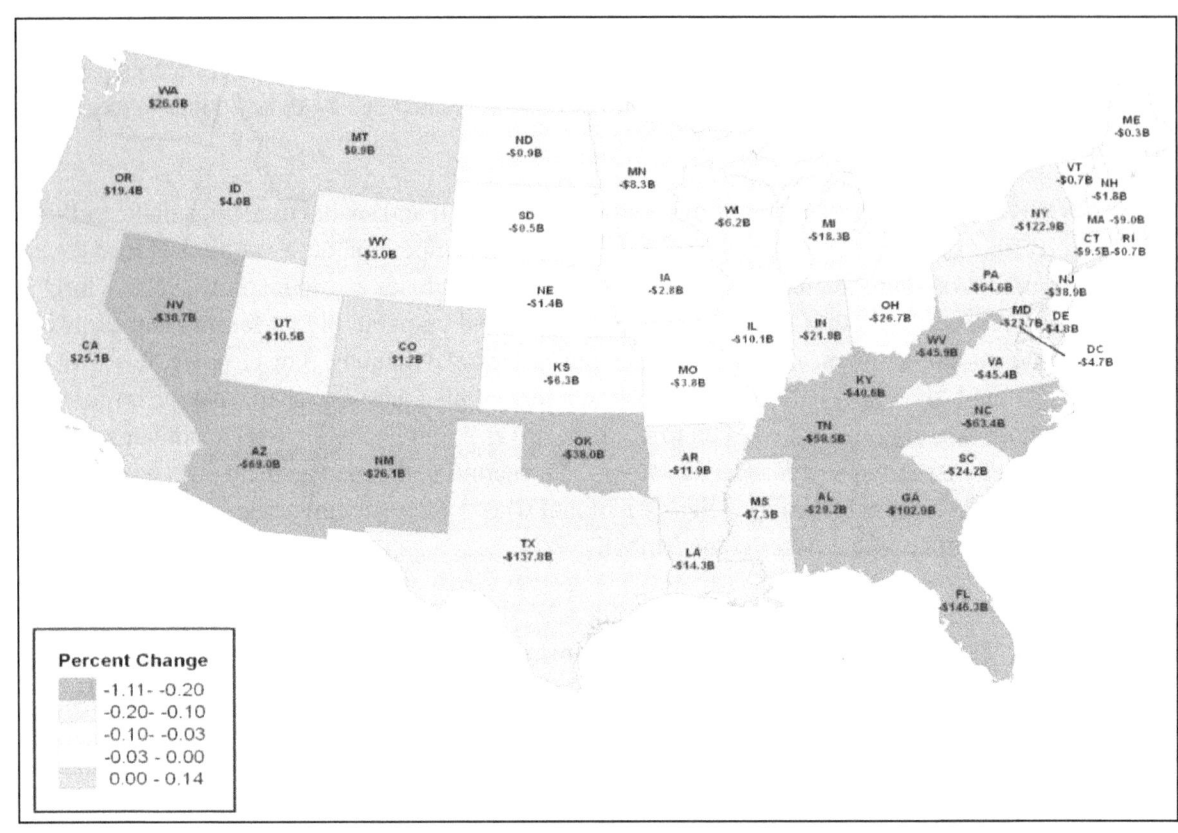

Figure 4.3. Quantified risk to state GDP ($B) from 2010 to 2050 due to uncertainty in precipitation predictions. From Backus et al. (2010).

Backus et al. (2010) show that uncertainty in climate predictions can translate into drastically different risks and consequences from state to state, meaning that the spatial distribution of risk is not necessarily a function of the distribution of the climatic stress. For multi-sector systems, the scale of risk, uncertainty, and vulnerability are as much of a function of the inter-sector technologies and dynamics as they are of the climate stress itself. As long as the system being analyzed is multi-sector in nature, the potential exists for the risks and vulnerabilities to manifest at spatial and temporal scales far beyond the original scales being considered.

4.2.4 Characteristic 4 – Low-Probability, High-Consequence Events

Low-probability, high-consequence events can be thought of as "threshold" events that occur when a key system is stressed beyond its breaking point. Prior to that time there may be little or no consequence, whereas afterwards the consequence may be catastrophic. The event in this case is the pivotal event that defines the consequence since a low-probability, high-consequence event can result from both long-term chronic stress as well as short-term acute stress. For example, a dam can fail from concrete weakening over many years or from a major flood event that breaches and destroys the dam. The pivotal event in this case is the dam breaking while the consequence is the destruction from the resulting flood.

Important lessons pertaining to this characteristic can be found outside the physical sciences in the economics literature. Economic assessments of the effects of climate change are cross-sectorial by definition since climate stressors must be translated across sectors into economic consequences. As a result, the economic community has put more emphasis on defining and characterizing cross-sector risk and uncertainty than probably any other discipline.

As part of a written symposium titled "Fat Tails and the Economics of Climate Change" in the summer 2011 edition of *Review of Environmental Economics and Policy*, Weitzman (2011) presents five exhibits that highlight and illustrate the key structural uncertainties associated with modeling and assessing the effects of climate change and their corresponding economic consequences. The fundamental lesson is that uncertainty (and by extension, risk and vulnerability) within climate impact assessment is multi-layered and complex, which is embodied in the summary of his paper (Weitzman 2011):

> "To summarize, the economics of climate change consists of a very long chain of tenuous inferences fraught with big uncertainties in every link: beginning with unknown base-case GHG emissions; then compounded by big uncertainties about how available policies and policy levers will transfer into actual GHG emissions; compounded by big uncertainties about how GHG flow emissions accumulate via the carbon cycle into GHG stock concentrations; compounded by big uncertainties about how and when GHG stock concentrations translate into global average temperature changes; compounded by big uncertainties about how global average temperature changes decompose into specific changes in regional weather patterns; compounded by big uncertainties about how adaptations to, and mitigations of, climate-change damages at a regional level are translated into regional utility changes via an appropriate damages function; compounded by big uncertainties about how future regional utility changes are aggregated into a worldwide utility function and what should be its overall degree of risk aversion; compounded by big uncertainties about what discount rate should be used to convert everything into expected-present-discounted values. The result of this lengthy cascading of big uncertainties is a reduced form of truly extraordinary uncertainty about the aggregate welfare impacts of catastrophic climate change, which mathematically is represented by a PDF [probability density function] that is spread out and heavy with probability in the tails."

In that same symposium, Nordhaus (2011) addresses the concepts of "fat tails" and what Weitzman called in earlier work (Weitzman 2009) the "dismal theorem." The dismal theorem is based on the premise that due to our limited understanding of the structure of uncertainty and societal preferences, the expected loss from climate change becomes infinite and standard economic analysis cannot be applied. In other words, *because of the uncertainty in the uncertainty, the likelihood of catastrophic events is not trivial and thus the exposure to damage (and risk) is virtually unlimited.*

Fat tailed distributions (sometimes called long-tailed distributions) are distributions where the probability of high-consequence events declines slowly, relative to "thin tailed" distributions like the Normal distribution (Weitzman 2011). The power-law (i.e., Pareto) distribution is an example of a fat tailed distribution. The implications of this are illustrated by Nordhaus (2011), where he uses earthquake records as an example. Over the last 200 years, the recurrence interval for an earthquake of any size follows a Pareto distribution. If earthquakes were normally distributed, then an earthquake as large as the March 2011 earthquake in Japan would occur once every 10 trillion years or so. However, using a fat tailed distribution like the Pareto (which earthquake records fit very nicely), the recurrence interval drops to once every 100 years or so. Considered from the return interval perspective, an earthquake that is described using a Normal distribution with a return interval of 100 years would be only *slightly larger* than the historical average (i.e., much smaller than the March 2011 event), even though it is the same number of standard deviations away from its mean as the March 2011 earthquake was from its Pareto mean.

Fat tailed distributions appear frequently in climate change impact analysis; a perfect example is the often referenced figure by Hegerl et al. (2007), recreated here as Figure 4.4. The figure shows the climate sensitivity predicted by nine different studies and the uncertainty contained in those predictions. Climate sensitivity refers to the projected temperature increase in response to a doubling of GHG concentrations. The colored lines in the lower half of the plot show the 5–95% confidence interval for its like-color curve in the upper half, with the mean of the prediction shown by the placement of the dot. All of the curves in the figure are making a statement that fat tailed uncertainty is inherent in predictions of climate sensitivity. The fact that the mean of each curve is greater than its most likely value (the value at the peak of each curve) is evidence of this. It is the tails of the distributions that skew the mean to be greater than the most likely estimate.

We can use this example to place a mathematical descriptor on fat tailed distributions and extreme events. Referring back to the definition of risk, the risk associated with climate sensitivity is the consequence of a given temperature rise multiplied by the probability that the rise will occur. Thus, the consequence, C, is a function of the temperature, T, while the temperature is a function of the probability, P (i.e., uncertainty). If the rate of change of consequence with respect to temperature is greater than the rate of change in temperature with respect to probability, the calculation of risk will be skewed toward the extreme events. In other words, if $\Delta C/\Delta T > \Delta T/\Delta P$, then the focus with regards to estimating risk should fall on high-consequence, low-likelihood events. This concept can be applied to any stressor or set of stressors assuming that the consequence can be adequately evaluated.

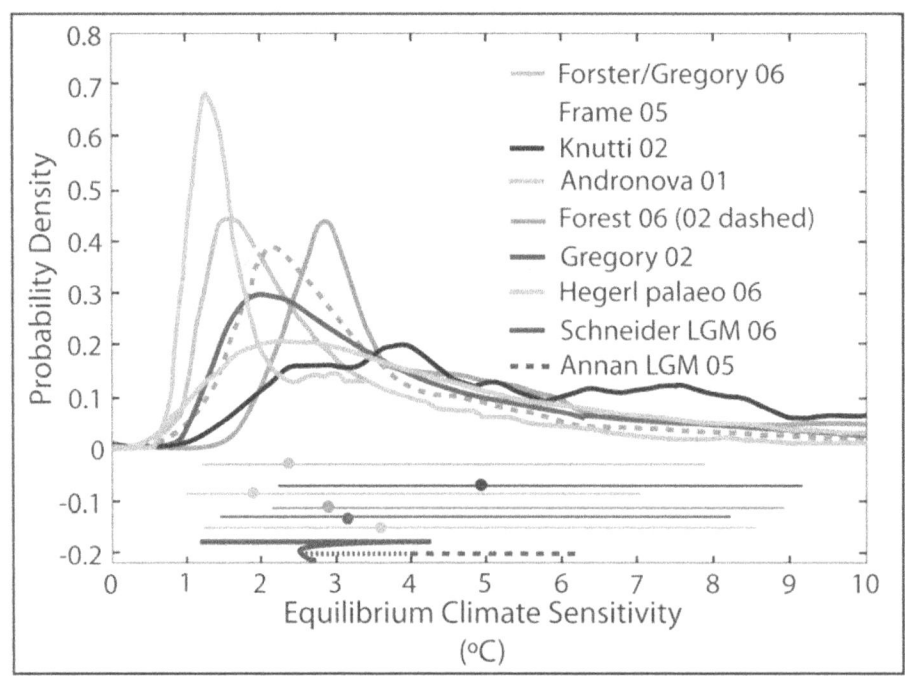

Figure 4.4. Climate sensitivity from Hegerl et al. (2007).

More evidence of the importance of low-probability, high-consequence events can be found by observing that most of our infrastructure and inter-sectorial technologies were built to withstand the majority of anticipated futures but not all future events. At some point, decisions are made (usually involving cost-benefit analysis) to limit the integrity and resilience of these systems to withstand impacts of a certain exceedance probability or greater. A clear example of this is the concept of the 100-year flood plain that is used to define homes that have a greater probability of flood damage. However, several flood events in recent years have exceeded the historical record and thus cannot be easily assigned likelihoods or recurrence intervals. For the most part, our infrastructure is built to withstand 100-year flood events, so the consequences of these events are relatively low. However, when events exceed our ability to withstand them, the consequence of those events can suddenly rise, as when the dike broke during Hurricane Katrina. In other words, the consequence of an event is only really experienced when the inter-sector dynamics are extended beyond the historical variability for which they were designed. It is then that risks and vulnerabilities are felt across sector interfaces.

Adding to this uncertainty is the fact that estimating the risks, uncertainties, and vulnerabilities of extreme events is difficult and problematic. While Nordhaus (2011) and Weitzman (2011) provide evidence that the effects of climate change are fat tailed and thus heavily influenced by extreme events, they also stress that it is very difficult to estimate the likelihood of extreme events from observed data when the distributions have fat tails. A flood event that exceeds the historical record, as discussed above, is an example of this. Furthermore, it is extremely difficult to determine if extreme events such as the Texas drought or Hurricane Katrina are a consequence of climate change, a part of the natural variability, or both. For instance, taken individually, the high temperature and the lack of rain during the Texas drought, as depicted in Figure 3.1, are each extreme events since they both lie far from their respective historical means, but it is the two in combination that creates an event beyond any on record. However, even after the fact it is difficult to say that the extreme nature of the drought was due to climate change or even to determine the degree to which climate change worsened the drought. Because there is only one data point, estimating the probability of potential future consequences and risk becomes difficult at best.

4.3 Summary

While the four characteristics are presented above as separate entities, they are intimately linked and in many ways describe a single, over-arching quality of risk, uncertainty, and vulnerability for cross-sector systems as highly complex concepts with multiple layers of feedback and far reaching consequences. Primary and secondary dynamics and feedback from uncertainty in the effect of decisions on adaptation and mitigation mean that cross-sector risk, uncertainty, and vulnerability are more broadly defined than when applied to single sectors or primary dynamics. Delays, feedbacks, human response and behavior, and changing operational landscapes can amplify and attenuate them in dramatic and unintuitive ways. The nature of multi-sector systems themselves means that the associated risks, uncertainties, and vulnerabilities reach across large scales of time and space that create interdependencies and secondary dynamics that otherwise would not exist. This can also lead to a decoupling between the spatial and temporal scales of climate effects and that of the risks and vulnerabilities. Finally, risk, uncertainty, and vulnerability in cross-sector systems is dominated by low-probability, high-consequence events, mainly due to the built-in resilience of existing technology and infrastructure to higher probability, lower consequence events.

Thus, decision making in the face of climate change and EWL system interactions must consider primary and secondary dynamics, human behavior and response, potential changes in the operational landscape, and the nature and magnitude of low-probability events. In addition, decision makers and analysts must understand how uncertainty in those considerations could amplify or attenuate the risks as well as how the inter-sector dynamics themselves could change as a function of climate change, our response to climate change, or both. This task is immensely difficult and in some ways insurmountable. However, by being cognizant of these characteristics and identifying their influence, decision makers and analysts can better weigh the strategic and practical tradeoffs of different decisions. Simple awareness of these characteristics also means that monitoring and assessment mechanisms can be set up to try to anticipate and avoid or correct unintended consequences, and that more effective emergency planning and response can be put in place by understanding the sources of those consequences.

Because uncertainty will never be eliminated, the goal must be in making decisions under uncertainty and allowing decision makers to adequately answer the questions: (1) What are the effects of climate change; and (2) What are the effects of our responses? While reducing uncertainty is a worthy objective, within a cross-sector world, it may be secondary to our need to identify and understand sector-to-sector dynamics, since it is the dynamics themselves that influence risk, uncertainty, and vulnerability in cross-sector systems the most. Based on the discussion above, the biggest gains will come by increasing our understanding of human response and behavior to climate change and decisions concerning climate change, identifying the trigger points where low-probability events in one sector can become high-consequence events in other sectors, and in identifying and understanding the amplification, attenuation, and feedback mechanisms that create unintended and unanticipated consequences.

4.4 References

Alpizar F, F Carlsson, and M Naranjo. 2011. "The Effect of Ambiguous Risk, and Coordination on Farmers' Adaptation to Climate Change - A Framed Field Experiment." *Ecological Economics* 70(12): 2317-2326.

Backus GA, TS Lowry, and DE Warren. 2012. "The near-term risk of climate uncertainty among the U.S. states." *Climatic Change*, in press.

Barabasi A-L. 2005. "The Origin of Bursts and Heavy Tails in Human Dynamics." *Nature* 435(7039): 207-211, DOI: 10.1038/nature03459.

Blarer A and M Doebeli. 1999. "Resonance Effects and Outbreaks in Ecological Time Series." *Ecology Letters* 2(3):167-177.

Brown R, R Ashley, and M Farrelly. 2011. "Political and Professional Agency Entrapment: An Agenda for Urban Water Research." *Water Resources Management* 25(15):4037-4050, DOI: 10.1007/s11269-011-9886-y.

Eckmann J-P, E Moses, and D Sergi. 2004. "Entropy of Dialogues Creates Coherent Structures in E-mail Traffic." *Proceedings of the National Academy of Science of the United States of America* 101(40): 14333-14337, DOI: 10.1073/pnas.0405728101.

Falloon P and R Betts. 2010. "Climate Impacts on European Agriculture and Water Management in the Context of Adaptation and Mitigation-The Importance of an Integrated Approach." *Science of the Total Environment* 408(23):5667-5687.

Freibauer A, M Rounsevell, P Smith, and J Verhagen. 2004. "Carbon Sequestration in the Agricultural Soils of Europe." *Geoderma* 122(1):1-23.

Ghoulmie F, R Cont, and J-P Nadal. 2005. "Heterogeneity and Feedback in an Agent-Based Market Model." *Journal of Physics-Condensed Matter* 17(14):S1259-S1268, DOI: 10.1088/0953-8984/17/14/015.

Greenman JV, and TG Benton. 2003. "The Amplification of Environmental Noise in Population Models: Causes and Consequences." *The American Naturalist* 161(2):225-239, DOI: 10.1086/345784.

Greenman JV and TG Benton. 2005. "The Frequency Spectrum of Structured Discrete Time Population Models: Its Properties and Their Ecological Implications." *Oikos* 110(2):369-389, DOI: 10.1111/j.0030-1299.2005.13652.x.

Greenman JV and TG Benton. 2005. "The Impact of Environmental Fluctuations on Structured Discrete Time Population Models: Resonance, Synchrony and Threshold Behaviour." *Theoretical Population Biology* 68(4):217-235, DOI: 10.1016/j.tpb.2005.06.007.

Hegerl GC, FW Zwiers, P Braconnot, NP Gillett, Y Luo, JA Marengo Orsini, N Nicholls, JE Penner, and PA Stott. 2007. "Understanding and Attributing Climate Change." Chapter 9 in *Climate Change 2007:The Physical Science Basis. Contribuion of Working Group I to the Fourth Assessment Report of the Intergovernmental Panel on Climate Change.* Cambridge University Press, Cambridge, United Kingdom and New York, New York, USA.

Hoff H. 2009. "Global Water Resources and Their Management." *Current Opinion in Environmental Sustainability* 1(2):141-147, DOI:10.1016/j.cosust.2009.10.001.

Hunt A and P Watkiss. 2011. "Climate Change Impacts and Adaptation in Cities: A review of the Literature." *Climatic Change* 104(1):13-49, DOI:10.1007/s10584-010-9975-6.

Jo H-H, E Moon, and K Kaski. 2012. "Optimized Reduction of Uncertainty in Bursty Human Dynamics." *Physical Review E* 85(1):016102/1-4, DOI:10.1103/PhysRevE.85.016102.

Kasperson RE. 2006. "Rerouting the Stakeholder Express." *Global Environmental Change: Human and Policy Dimensions* 16(4):320-322, DOI:10.1016/j.gloenvcha.2006.08.002.

Kasperson RE, O Renn, P Slovic, HS Brown, J Emel, R Goble, JX Kasperson, and S Ratick. 1988. "The Social Amplification of Risk: A Conceptual Framework." *Risk Analysis* 8(2):177-187, DOI:10.1111/j.1539-6924.1988.tb01168.x.

Koutsoyiannis D, C Makropoulos, A Langousis, S Baki, A Efstratiadis, A Christofides, G Karavokiros, and N Mamassis. 2009. "HESS Opinions: "Climate, Hydrology, Energy, Water: Recognizing Uncertainty and Seeking Sustainability." *Hydrology and Earth System Sciences* 13(2):247-257, DOI:10.5194/hess-13-247-2009.

Manning M, M Petit, D Easterling, J Murphy, A Patwardhan, HH Rogner, R Swart, and G Yohe, eds. 2004. *IPCC Workshop on Describing Scientific Uncertainties in Climate Change to Support Analysis of Risk and of Options.* Intergovernmental Panel on Climate Change, Boulder, Colorado.

Martin R. 2011. "The Local Geographies of the Financial Crisis: From the Housing Bubble to Economic Recession and Beyond." *Journal of Economic Geography* 11(4):587-618, DOI:10.1093/jeg/lbq024.

Mastrandrea MD, CB Field, TF Stocker, O Edenhofer, KL Ebi, DJ Frame, H Held, E Kriegler, KJ Mach, PR Matschoss, GK Plattner, GW Yohe, and FW Zwiers. 2010. Guidance Note for Lead Authors of the IPCC Fifth Assessment Report on Consistent Treatment of Uncertainties. Intergovernmental Panel on Climate Change, Geneva, Switzerland.

Miles EL, MM Elsner, JS Littell, LW Binder, and DP Lettenmaier. 2010. "Assessing Regional Impacts and Adaptation Strategies for Climate Change: The Washington Climate Change Impacts Assessment." *Climatic Change* 102(1-2):9-27, DOI:10.1007/s10584-010-9853-2.

Morgan, GM, H Dowlatabadi, M Henrion, D Keith, R Lempert, S McBride, M Small and T Wilbanks. 2009. CCSP, 2009: Best practice approaches for characterizing, communicating, and incorporating scientific uncertainty in decisionmaking, A Report by the U.S. Climate Change Science Program and the Subcommittee on Global Change Research Washington D.C., USA, National Oceanic and Atmospheric Administration.

Moss RH and SH Schneider. 2000. "Uncertainties in the IPCC TAR: Recommendations to Lead Authors for More Consistent Assessment and Reporting." In *Guidance Papers on the Cross Cutting Issues of the Third Assessment Report of the IPCC*, eds. R Pachauri, T Taniguchi, and K Tanaka, pp. 33-51. World Meteorological Organization, Geneva, Switzerland.

Moss RH and G Yohe. 2011. Assessing and Communicating Confidence Levels and Uncertainties in the Main Conclusions of the NCA 2013 Report: Guidance for Authors and Contributors. National Climate Assessment Development and Advisory Committee (NCADAC), http://usgcrp.gov. Washington D.C.

Newell B, DM Marsh, and D Sharma. 2011. "Enhancing the Resilience of the Australian National Electricity Market: Taking a Systems Approach in Policy Development." *Ecology and Society* 16(2):15.

REMI - Regional Economics Models, Inc. 2009. Variable description for Production Cost, REMI PI+, v.1.1, March 24, 2009 build, 51-region, 70-sector model, Regional Economics Models, Inc., Amherst, Massachusetts.

Renn O. 2011. "The Social Amplification/Attenuation of Risk Framework: Application to Climate Change." *Wiley Interdisciplinary Reviews: Climate Change* 2(2):154-169, DOI:10.1002/wcc.99.

Scott CA and MJ Pasqualetti. 2010. "Energy and Water Resources Scarcity: Critical Infrastructure for Growth and Economic Development in Arizona and Sonora." *Natural Resources Journal* 50(3):645-682.

Scott CA, SA Pierce, MJ Pasqualetti, AL Jones, BE Montz, and JH Hoover. 2011. "Policy and Institutional Dimensions of the Water-Energy Nexus." *Energy Policy* 39(10):6622-6630, DOI:10.1016/j.enpol.2011.08.013.

Socolow RH. 2011. "High-Consequence Outcomes and Internal Disagreements: Tell Us More, Please." *Climatic Change* 108:775-790, DOI:10.1007/s10584-011-0187-5.

Tidwell VC, PH Kobos, LA Malczynski, G Klise, and CR Castillo. 2012. "Exploring the water-thermoelectric power nexus." *Journal of Water Planning and Management,* in press.

van Pelt SC and RJ Swart. 2011. "Climate Change Risk Management in Transnational River Basins: The Rhine." *Water Resources Management* 25(14):3837-3861, DOI:10.1007/s11269-011-9891-1.

Webber ME. 2011. "The Nexus of Energy and Water in the United States." In *Physics of Sustainable Energy II: Using Energy Efficiently And Producing It Renewably,* eds. D Hafemeister, D Kammen, BG Levi, and P Schwartz, pp. 84-106. March 5-6, 2011, Berkeley, California, American Institute of Physics, Melville, New York, DOI: 10.1063/1.3653847.

Weil HB. 2010. "Why Markets Make Mistakes." *Kybernetes* 39(9-10):1429-1451, DOI:10.1108/03684921011081114.

Weitzman ML. 2009. "On Modeling and Interpreting the Economics of Catastrophic Climate Change." *Review of Economics and Statistics* 91(1):1-19, DOI:10.1162/rest.91.1.1.

Weitzman ML. 2011. "Fat-Tailed Uncertainty in the Economics of Catastrophic Climate Change." *Review of Environmental Economics and Policy* 5(2):275-292, DOI:10.1093/reep/rer006.

Yohe G and M Oppenheimer. 2011. "Evaluation, Characterization, and Communication of Uncertainty by the Intergovernmental Panel on Climate Change-An Introductory Essay." *Climatic Change* 108(4):629-639, DOI:10.1007/s10584-011-0176-8.

5.0 Climate Mitigation and Adaption at the Energy-Water-Land Interfaces

Energy-water-land (EWL) linkages are rudimentary to mitigation and adaptation. Many mitigation and adaptation options tie directly into one or more of the EWL sectors, and are therefore tied into the EWL interfaces. Understanding the EWL nexus is therefore central to the effective design, selection, implementation, and monitoring of adaptation and mitigation.

Almost all mitigation options lie within either the energy or land sectors. Mitigation reduces or sequesters emissions arising from the supply and demand for energy and land (e.g., substituting renewable technologies for fossil fuel generation; preventing deforestation). As such, mitigation options are affected by EWL relationships. Some are directly affected because they create demands for water or land endowments (e.g., CCS or bioenergy). Others are indirectly affected as alternatives for energy supply and production or land-use. Energy, water, and land resources are also vulnerable to climate change. Adaptation measures are responses to climate change related risks. Therefore, adaptation options designed to reduce vulnerability to climate impacts in one EWL sector affect, and are affected by, EWL linkages. Some adaptation measures reduce demands on EWL endowments (e.g., water-use efficiency), while others may increase them (e.g., desalinization).

Should mitigation and adaptation be viewed differently because of the EWL interfaces? Ties to the nexus create vulnerabilities and opportunities for mitigation and adaptation options. EWL vulnerabilities are risks, and therefore costs, which vary by mitigation and adaptation option and location. Thus, EWL risk effects could change the relative appeal of mitigation and adaptation options. Similarly, mitigation and adaptation options that reduce EWL stress are creating additional value in the form of avoided risk. These EWL ties are opportunities for managing EWL risks—(1) for vulnerable individual mitigation and adaptation options by reducing EWL demands, enhancing endowments, or improving the efficiency of linkage technology, and (2) across mitigation and adaptation options where increased deployment of one option could reduce EWL risks for others.

Furthermore, because both mitigation and adaptation tie into EWL interfaces, they are also linked to each other through the nexus. Therefore, there are coordination challenges and opportunities. Overall, adaptation and mitigation planning should consider EWL relationships. Doing so requires integrated analyses to better understand relationships and trade-offs, manage risks, and exploit opportunities.

Understanding mitigation and adaptation relationships to the EWL interfaces facilitates not only the evaluation of the net impact of individual mitigation or adaptation measures, but also the compound effects of concurrent implementation, either intentionally or as an outcome of the uncoordinated actions of independent parties. These compound effects may not always have positive synergies, and ignoring or failing to identify potentially negative interactions runs the risk of undermining the original policy goals (Moser 2012).

In section 5.2, we characterize relationships between mitigation and adaptation options and EWL linkages. In section 5.3, we discuss mitigation and adaptation decision-making vulnerabilities, opportunities, and coordination in light of their EWL relationships.

5.1 Mitigation, Adaptation, and EWL Linkages

This section illustrates relationships between mitigation and adaptation options and the EWL interfaces. In particular, this section illustrates relative differences in the relationships across options, which represent differences in EWL risk exposure as well as EWL risk management opportunities.

5.1.1 Mitigation

Figure 5.1 summarizes water withdrawal and consumption ranges per megawatt-hour by electricity generation plant type and cooling system (Macknick et al. 2011). There are large differences in water use impacts across generation options, as well as between cooling system alternatives for a single type of generation. Averyt et al. (2011) note:

> "Water withdrawals per megawatt-hour (mWh) can range from almost zero for a solar photovoltaic, wind, or dry-cooled natural gas plant, to hundreds of gallons for an efficient plant using recirculating cooling, to tens of thousands of gallons for a nuclear or coal plant using once-through cooling. Water consumption per mWh can similarly range from almost zero for solar, wind, or gas plants using dry cooling to around 1,000 gallons for coal, oil, or concentrating solar power (cSP) with recirculating cooling. How much water a specific plant uses reflects its efficiency and age, and how much the plant is used, along with local humidity, air temperature, and water temperature."

Increased deployment of generation with higher water-use impacts implies increasing EWL vulnerability, with differing impacts associated with high withdrawal versus high consumption (Averyt et al. 2011). Conversely, deployment of lower water-use generation represents an opportunity for managing overall EWL risk.

Particularly relevant to climate mitigation are the EWL risks associated with low-carbon electricity generation. Nuclear and coal with CCS are projected to be a potentially substantial part of a future decarbonized energy system (e.g., Clarke et al. 2007; Fisher et al. 2007; USEPA 2010). However, both are also potentially water intensive and therefore vulnerable to conflicts across the EWL interfaces. For example, coal-fired power plants with evaporative cooling fitted with CCS would consume twice as much water per unit of electricity generated as a non-CCS coal-fired facility (Zhai et al. 2011). For a plant using dry cooling (air-cooled condensers), water use would be reduced by about 80% without CCS and about 40% with CCS—but at approximately triple the capital cost.. Alternatively, renewable generation and combined cycle gas and coal have relatively modest water withdrawals (see also EPRI 2011). Overall, EWL vulnerability is an important factor to weigh in considering alternative generation options and cooling systems.

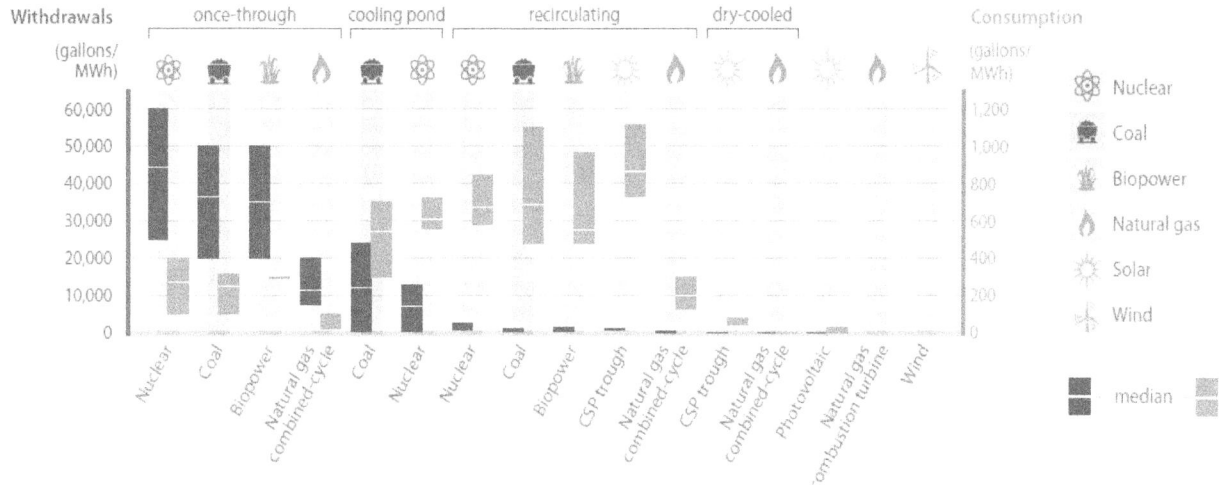

Figure 5.1. Electricity generation water use by fuel and cooling technology. Ranges reflect minimum and maximum water-use values for selected technologies. Horizontal lines within rectangles indicate median values. (Source: Macknick et al. (2011) as reported in Averyt et al. (2011))

Similar EWL relative vulnerability characterizations exist for the relationship between climate mitigation options to the nexus within the energy-land and land-water linkages, as well as for non-electric energy (industrial, residential, transportation). For instance, in the case of the land-water linkages, some potential bioenergy feedstocks (e.g., conventional and energy crops) are land intensive and therefore exposed to EWL risk in the form of competition for land resources (e.g., with food and feed commodities), water resources (e.g., with municipalities, power, and industry), and climate-related effects on the productivity and availability of land and water endowments. Other forms of non-bioenergy land-based GHG mitigation are also tied intimately into the interfaces with, in particular, implications for land availability, land management, and water resource quantity and quality (e.g., afforestation, forest management, agricultural soil, and fertilizer management). For examples of land allocation and management implications, see Calvin et al. (2009), Golub et al. (2009), and Rose and Sohngen (2011).

5.1.2 Adaptation

Climate adaptation measures are responses to current or projected climate change impacts that reduce potential losses or enhance opportunities. Each of the EWL sectors could be affected by climate change. Table 5.1 provides a high-level characterization from the U.S. Global Change Research Program of potential climate change impacts that could affect the U.S. energy, water, and land sectors. Specific impacts, of course, vary substantially by location.

Adaptation responses to the types of impacts listed in Table 5.1 can be autonomous or non-autonomous, where the former refers to endogenous responses of individuals given current knowledge and technologies (e.g., farmer changes in crops and management, household changes in cooling and heating), and the latter refers to strategic policy and institutional decisions to expand knowledge and technology (e.g., changes in water infrastructure, drought resistant cultivar research and development).

Adaptation options to any specific EWL impact, such as those discussed in section 5.3, will have benefits for the specific EWL sector affected. Implications for the nexus will vary by impact and adaptation option. These relative differences in nexus implications are opportunities for managing risks and exploiting synergies across the nexus and between adaptation and mitigation. Section 5.3 focuses on this complex coordination, and current understanding.

Table 5.1. Key Messages Regarding Climate Change Impacts on U.S. Energy Supply and Use, Water Resources, Agriculture, and Terrestrial Ecosystems (Source: Karl et al. 2009)

Energy Supply and Use
• Warming will be accompanied by decreases in demand for heating energy and increases in demand for cooling energy. The latter significantly increase electricity use and create higher peak demand in most regions.
• Energy production is likely to be constrained by rising temperatures and limited water supplies in many regions.
• Energy production and delivery systems are exposed to sea-level rise and extreme weather events in vulnerable regions.
• Climate change is likely to affect some renewable energy sources across the nation, such as hydropower production in regions subject to changing patterns of precipitation or snowmelt.

Water resources
• Climate change has already altered, and will continue to alter, the water cycle, affecting where, when, and how much water is available for all uses.
• Floods and droughts are likely to become more common and more intense as regional and seasonal precipitation patterns change, and rainfall becomes more concentrated into heavy events (with longer, hotter dry periods in between).
• Precipitation and runoff are likely to increase in the Northeast and Midwest in winter and spring, and decrease in the West, especially the Southwest, in spring and summer.
• In areas where snowpack dominates, the timing of runoff will continue to shift to earlier in the spring and flows will be lower in late summer.
• Surface water quality and groundwater quantity will be affected by a changing climate.
• Climate change will place additional burdens on already stressed water systems.
• The past century is no longer a reasonable guide to the future for water management.

Agriculture
• Many crops show positive responses to elevated carbon dioxide and low levels of warming, but higher levels of warming often negatively affect growth and yields.
• Extreme events such as heavy downpours and droughts are likely to reduce crop yields because excesses or deficits of water have negative impacts on plant growth.
• Weeds, diseases, and insect pests benefit from warming, and weeds also benefit from a higher carbon dioxide concentration, increasing stress on crop plants and requiring more attention to pest and weed control.
• Forage quality in pastures and rangelands generally declines with increasing carbon dioxide concentration because of the effects on plant nitrogen and protein content, reducing the land's ability to supply adequate livestock feed.
• Increased heat, disease, and weather extremes are likely to reduce livestock productivity.

Ecosystems*
• Ecosystem processes, such as those that control growth and decomposition, have been affected by climate change.
• Large-scale shifts have occurred in the ranges of species and the timing of the seasons and animal migration, and are very likely to continue.
• Fires, insect pests, disease pathogens, and invasive weed species have increased, and these trends are likely to continue.
• Deserts and drylands are likely to become hotter and drier, feeding a self-reinforcing cycle of invasive plants, fire, and erosion.
• The habitats of some mountain species and coldwater fish, such as salmon and trout, are very likely to

> contract in response to warming.
> - Some of the benefits ecosystems provide to society will be threatened by climate change, while others will be enhanced.

* Only terrestrial ecosystem points included. For ocean ecosystem points, see Karl et al. (2009).

A number of the case studies in this report illustrate relationships between mitigation and adaptation options and the EWL nexus. For instance, section 3.3.5 provides examples of regional electricity generation alternatives and EWL relationships and implications in the context of renewable generation and climate policies (e.g., renewables in California, and coal-fired generation with CCS in the Southeast). Similarly, section 3.2.2 discusses Texas drought water scarcity adaptation responses in the form of water transfers, potential changes in thermoelectric power cooling systems, and drought programs. These actions have implications not only for Texas municipal water supplies, but also overall stress on the nexus and the vulnerabilities of all activities tied to the nexus.

5.2 Decision-making Vulnerabilities, Opportunities, and Coordination for Mitigation and Adaptation at EWL Interfaces

Many actions to reduce emissions in the energy and land-use sectors have cross-sectoral impacts. So too do sector-specific actions to adapt to climate change impacts in the energy, land, and water sectors (see, e.g., Rice et al. 2012; Bedsworth and Hanak 2008; Vine 2008; CEC 2005).

As shown in Table 5.2, some sector-specific mitigation and adaptation measures have the potential to provide synergistic "win-win" opportunities to enhance climate mitigation or adaptation objectives across one or more other sectors in the nexus. However, other measures may have negative impacts on mitigation or adaptation potential in other sectors. Such cross-sectoral impacts can carry substantial risks of inadvertently diminishing climate mitigation and/or adaptation objectives.

For example, the mitigation measure of switching from coal to natural-gas-fired electricity generation not only reduces the emissions associated with power generation, it also improves adaptive capacity to water stress because it reduces water use for thermoelectric cooling (gas-fired plants have much reduced needs for cooling water compared to coal-fired plants). At the same time, however, this fuel switching may also reduce adaptive capacity to water stress because natural gas extraction techniques both consume water and have the potential to contaminate water supplies. In addition, land-based ecosystems may be stressed due to the potential for habitat fragmentation from gas exploration and extraction (e.g., wildlife migration).

Incorporating consideration of such cross-sectoral interactions in planning and policy could affect sectoral decisions and overall portfolios for climate mitigation and adaptation. To varying degrees, a combination of careful planning and targeted research and development can enhance synergies and reduce the risks of negative impacts. In the case of fuel switching from coal to gas, research and development (R&D) in gas drilling/stimulation technologies and water re-use may result in a reduction of negative impacts on water supplies and enhance the synergies across the EWL interface.

It is important to keep in mind that Table 5.2 only reflects physical synergies and trade-offs. There are, of course, economic trade-offs as well in the form of technology costs, revenues, prices, and societal welfare, and perhaps tradeoffs with other policy objectives, such as energy security.

Table 5.2. Cross-Sectoral Impacts of Climate Mitigation and Adaptation Measures on Climate Objectives: Examples of Synergies, Risks, and Response Options Across the Energy-Water-Land Nexus

Sector-specific Mitigation Measures		Positive Implications for Mitigation	Potential for Synergies across the EWL Nexus	Risk of Negative Implications across the EWL Nexus	Options for Enhancing Synergies and Reducing Risks	Key References
Energy	**Increase energy efficiency**	• Reduce per capita emissions	• W, L: Improve ability to adapt to diminished water and/or land resources through reduced demand for electricity and transportation fuels • E, L: Potential to reduce or delay need for new fossil-based energy resources and associated land use	• E: Costs for service suppliers could rise depending on the approach/technology for energy efficiency improvements, subsidies, and cost/benefit allocation between consumers and service suppliers	• Improve/streamline financial incentives for energy consumers and suppliers of energy technologies	*CEC 2005*
	Switch from coal to natural gas fuels	• Reduce per capita emissions	• W: Improve ability to adapt to diminished water resources by decreasing water per kilowatt-hour	• L: Reduce wildlife migration options via habitat fragmentation from natural gas extraction activities • W: Reduce ability to adapt to diminished water resources by increasing water demand for gas well stimulation and potential pollution of potable supplies	• R&D in drilling/stimulation technologies and water re-use to reduce risk of negative impacts of hydrofracking on water supplies.	*Climate Decision Making Illustration 3.2 (see section 3.3.5)*

5.6

Sector-specific Mitigation Measures	Positive Implications for Mitigation	Potential for Synergies across the EWL Nexus	Risk of Negative Implications across the EWL Nexus	Options for Enhancing Synergies and Reducing Risks	Key References
Expand CCS to fossil-fueled power plants	• Reduce per capita emissions		• W: Reduce ability to adapt to diminished water resources by increasing water per kilowatt-hour • L: Limit utility of land above sequestration sites due to seismicity and leakage potential	• R&D in water efficiency of CCS technologies • Siting of CCS plants in current and projected low water stress basins	*Illustration 3.1 (see section 3.3.2)*
Reduce reliance on fossil fuels through....					
Expansion of nuclear power	• Reduce per capita emissions		• W: For plants with cooling towers, high water withdrawal and consumption reduce resilience to diminished freshwater supplies • W: For plants with once-through cooling, elevated effluent temperatures because of climate driven increases in both fresh and saline water temperatures can reduce resilience of aquatic and marine ecosystems	• R&D and implementation of wastewater, low-water use, and dry cooling technologies • Siting of nuclear plants in current and projected low water stress basins • Ongoing environmental assessment of thermal pollution and impacts, incorporating projections of a future local climate	*Illustrations 2.1 and 2.2 (see section 2.2.2.1)*

5.7

Sector-specific Mitigation Measures	Positive Implications for Mitigation	Potential for Synergies across the EWL Nexus	Risk of Negative Implications across the EWL Nexus	Options for Enhancing Synergies and Reducing Risks	Key References
Expansion of wind	• Reduce per capita emissions	• W: Increased resilience to current and projected drought if thermoelectric cooling needs decrease	• L: Reduce wildlife migration options via habitat fragmentation, and land-use conflicts from dedicated lands for wind farms and transmission • L: Increased land-use and associated land-use conflict if variability of production requires base-load expansion and/or development of expanded storage • E: Need for wind firming resources could increase electricity costs	• Reducing tradeoffs by considering wildlife habitat and corridors in planning and siting, and enhancing synergies through siting	*Illustration 2.4 (see section 2.2.2.3)*
Expansion of large-scale PV technologies	• Reduce per capita emissions	• W: Increased resilience to current and projected drought	• L: Reduce wildlife migration options via habitat fragmentation from dedicated lands for solar farms and transmission • L: Increased land use and associated land-use conflict if variability of production requires base-load expansion and/or development of expanded storage	• Reducing tradeoffs by considering wildlife habitat and corridors in planning and siting, and enhancing synergies through siting in optimal-wind and water stressed regions	*Illustration 2.4 (see section 2.2.2.3)*

5.8

Sector-specific Mitigation Measures	Positive Implications for Mitigation	Potential for Synergies across the EWL Nexus	Risk of Negative Implications across the EWL Nexus	Options for Enhancing Synergies and Reducing Risks	Key References
Expansion of solar thermal technologies	• Reduce per capita emissions		• L: Reduce wildlife migration options via habitat fragmentation from dedicated lands for solar farms and transmission • W: For plants with cooling towers, high water withdrawal and consumption reduce resilience to diminished freshwater supplies	• Reducing tradeoffs by considering wildlife habitat and corridors in planning and siting • R&D and implementation of wastewater, low-water use, hybrid and dry cooling technologies • Siting of solar thermal plants in current and projected low water stress basins	*Illustrations 2.1 and 2.4 (see sections 2.2.2.1 and 2.2.2.3)*
Expansion of hydropower	• Reduce per capita emissions	• E: Renewable and dispatchable energy source for balancing non-dispatchable wind and solar production	L: Expanded hydropower could divert flows and/or create new reservoirs that could negatively impact sensitive ecosystems	• Development of adaptive reservoir management strategies • R&D to address minimum flow and water quality needs for ecosystems	
Invest in smart grid/demand response technologies	• Reduce per capita emissions	• E, L, W: Reduce need for new generation and transmission expansion potentially resulting in lower electricity costs, GHG emissions, land use, and water use			*Rice et al. 2012*

Sector-specific Mitigation Measures		Positive Implications for Mitigation	Potential for Synergies across the EWL Nexus	Risk of Negative Implications across the EWL Nexus	Options for Enhancing Synergies and Reducing Risks	*Key References*
Land	Expansion of biomass production for power generation and biofuels	• Reduce per capita emissions	• L, W: Expanded thinning of forests could reduce wildfire risk while improving water production	• W, L: Expanded cultivation of lands could increase stress on water resources and wildlife • L: Potential for increased GHG emissions from land clearing	• R&D to quantify and optimize linkages between vegetative cover and water production • R&D for sustainable biomass practices	*Illustration 2.1 (see section 2.2.2.1)*
	*Reforestation, afforestation	• Reduce net emissions through carbon sequestration	• L: Increase connectivity of habitat • W: Restoration of watershed function to denuded landscapes • E: Availability of material for biomass energy and timber production • L: Improve soil stabilization and reduce sediment loading to streams	• W: Enhance risk of increased water stress with high-water demanding tree species	• Choice of native and/or low-water use species • Application of best forest management practices	*Rice et al. 2012*

5.10

Sector-specific Adaptation Measures		Positive Implications for Sector-specific Adaptation	Potential for Synergies across the EWL nexus	Risk of Negative Implications across the EWL nexus	Options for Enhancing Synergies and Reducing Risks	Key References
Water	Increase freshwater use efficiency	• Reduce freshwater use per capita	• E, L: Increase availability of water for energy and agriculture • E: Reduced water use means reduced energy to capture, convey and treat the water, potentially reducing per capita emissions	• W: Resiliency gains from water use reductions may be lost if conserved water is used, e.g., to expand planted acreage	• Retire unused water rights to preserve water availability gains	*Vine 2008; CEC 2005*
	Switching from use of freshwater to wastewater	• Reduce freshwater use per capita	• E, L: Increase availability of freshwater and overall water supply for energy, agriculture, and other uses (e.g., municipal, ecosystems)	• E: Potential for increased energy requirements (i.e., increased GHG emissions) for treatment and pumping	• Using low-C energy sources for water treatment and delivery • Retire unused water rights to preserve water availability gains	*CEC 2005*
	Switch from once-through cooling to recirculating cooling of thermoelectric power plants	• Reducing water withdrawals and associated thermal pollution		• E, W: Increased risk of reducing availability of water for thermoelectric cooling and other demands (e.g., ag, muni) • E: Increased risk of reduced power generation efficiency because of increasing air and water temperatures	• R&D and implementation regarding more energy efficient hybrid and dry-cooling technologies	*Illustrations 2.1 and 2.2 (see section 2.2.2.1) CEC 2005*

Sector-specific Adaptation Measures	Positive Implications for Sector-specific Adaptation	Potential for Synergies across the EWL nexus	Risk of Negative Implications across the EWL nexus	Options for Enhancing Synergies and Reducing Risks	Key References
Switch from wet to dry cooling at thermoelectric power plants	• Reducing water use and associated thermal pollution		• E: Risk of reduced power generation efficiency because of increasing air temperatures • E: Increased energy costs due to higher capital and operating costs for dry cooling	• R&D to reduce costs of dry-cooling technologies	*Illustrations 2.1 and 2.2 (see section 2.2.2.1), CEC 2005*
Desalinization	• Increase brackish and freshwater supplies	• E, L: Increase availability of freshwater and overall water supply for energy, agriculture, and other uses (e.g., municipal, ecosystems)	• E: Potential for increased energy requirements (i.e., increased GHG emissions) • W: Enhanced risks for impacted marine ecosystems	• Use low-C energy sources for desalination • Ongoing environmental assessment of impacts, incorporating projections of a future local climate	*Illustration 2.3 (see section 2.2.2.2) CEC 2005*
New storage and conveyance of water to serve new demands	• Increase water supplies to meet demands	• Increase availability of freshwater and overall water supply for energy, agriculture, and other uses (e.g., municipal, ecosystems)	• E: Potential for increased energy requirements (i.e., increased GHG emissions) to move water • L, W: Potential to divert flows and/or create new reservoirs/infrastructure that could cause land use conflicts and negatively impact sensitive ecosystems		*Illustration 2.3 (see section 2.2.2.2)*

Sector-specific Adaptation Measures		Positive Implications for Sector-specific Adaptation	Potential for Synergies across the EWL nexus	Risk of Negative Implications across the EWL nexus	Options for Enhancing Synergies and Reducing Risks	Key References
	Watershed management	• Increase water supplies to meet demands	• Increase availability of freshwater and overall water supply for energy, agriculture, and other uses (e.g., municipal, ecosystems) • L, W: Reduce wildfire danger and improve water quality • L: Reduce flooding potential • L: Reduce need for CSO infrastructure upgrades	• L: Potential for increased land use conflicts if development is restricted in floodplains/riparian areas	• R&D for data collection and monitoring of stream flows, groundwater levels, ecosystem impacts	
Land	Switch to drought-tolerant crops	• Increased or maintained crop yields in drought vulnerable areas	• W: Reduced water demand	• L: Potential for reduced ability to maintain crop yields L: Shifts in ag land use/land cover could increase land use competition	• Retire unused water rights to preserve water availability gains	Rice et al. 2012
Energy	Increase transmission capacity to urban areas	• Reduce economic, social, and human health impacts due to power outages during heat waves	• E: Potential for reduced emissions if new transmission facilitates access of wind and solar to the grid	• L: Potential for land use conflicts for transmission lines	• R&D to address transmission needs and regulatory issues	Illustration 2.5 (see section 2.3)

5.13

Sector-specific Adaptation Measures		Positive Implications for Sector-specific Adaptation	Potential for Synergies across the EWL nexus	Risk of Negative Implications across the EWL nexus	Options for Enhancing Synergies and Reducing Risks	Key References
	Increase generation reserve resources in urban areas	• Reduce economic, social, and human health impacts due to power outages during heat waves		• E: Potential for increased emissions • L: Potential for siting conflicts over new power plants • W: Potential for water use conflicts over need for cooling water	• R&D/utilize/invest in dispatchable, clean urban storage resources, such as electric vehicles	*Rice et al. 2012*

* Also a sector-specific adaption measure.

5.3 References

Averyt K, J Fisher, A Huber-Lee, A Lewis, J Macknick, N Madden, J Rogers, and S Tellinghuisen. 2011. Freshwater use by U.S. power plants: Electricity's thirst for a precious resource, A report of the Energy and Water in a Warming World Initiative. Cambridge, MA: Union of Concerned Scientists. November. Bedsworth L, Hanak E (2008) Preparing California for a changing climate. PPIC, San Francisco.

CEC - California Energy Commission. 2005. California's Water-Energy Relationship. State of California, CEC-700-2005-011-SF.

Calvin K, J Edmonds, B Bond-Lamberty, L Clarke, SH Kim, P Kyle, SJ Smith, A Thomson, M Wise. 2009. "2.6: Limiting climate change to 450 ppm CO2 equivalent in the 21st century." *Energy Economics* 31(Supplemental 2):S107-S120.

Clarke L, J Edmonds, H Jacoby, H Pitcher, J Reilly, R Richels. 2007. Scenarios of Greenhouse Gas Emissions and Atmospheric Concentrations. Sub-report 2.1A of Synthesis and Assessment Product 2.1 by the U.S. Climate Change Science Program and the Subcommittee on Global Change Research. Department of Energy, Office of Biological & Environmental Research, Washington, DC., USA, 154 pp.

EPRI - Electric Power Research Institute. 2011. Water Use for Electricity Generation and Other Sectors: Recent Changes (1985-2005) and Future Projections (2005-2030). EPRI Technical Report 1023676, Electric Power Research Institute, November, pp 94.

Fisher BS, N Nakicenovic, K Alfsen, J Corfee Morlot, F de la Chesnaye, J-Ch Hourcade, K Jiang, M Kainuma, E La Rovere, A Matysek, A Rana, K Riahi, R Richels, S Rose, D van Vuuren, R Warren. 2007. Issues related to mitigation in the long term context, In Climate Change 2007: Mitigation. Contribution of Working Group III to the Fourth Assessment Report of the Inter-governmental Panel on Climate Change [eds. B Metz, OR Davidson, PR Bosch, R Dave, LA Meyer], Cambridge University Press, Cambridge, United Kingdom and New York, NY, USA.

Golub A, T Hertel, H-L Lee, S Rose, and B Sohngen. 2009. "The Opportunity Cost of Land Use and the Global Potential for Greenhouse Gas Mitigation in Agriculture and Forestry." *Resource and Energy Economics* 31:299-319.

Karl TR, JM Melillo, and TC Peterson, (eds.). 2009. Global Climate Change Impacts in the United States. Cambridge University Press.

Klein RJT, S Huq, F Denton, TE Downing, RG Richels, JB Robinson, FL Toth. 2007. 2007: Inter-relationships between adaptation and mitigation. Climate Change 2007: Impacts, Adaptation and Vulnerability. Contribution of Working Group II to the Fourth Assessment Report of the Intergovernmental Panel on Climate Change, eds. ML Parry, OF Canziani, JP Palutikof, PJ van der Linden and CE Hanson, Cambridge University Press, Cambridge, UK, 745-777.

MacKnick J, R Newmark, G Heath, and KC Hallet. 2011. A review of operational water consumption and withdrawal factors for electricity generating technologies. National Renewable Energy Laboratory, Golden, Colorado. Accessed at http://www.nrel.gov/docs/fy11osti/50900.pdf.

Moser SC. 2012. "Adaptation, mitigation, and their disharmonious discontents: an essay." *Climatic Change* 111(2):165-175, DOI: 10.1007/s10584-012-0398-4.

Rice JS, RH Moss, PJ Runci, KL Anderson, and EL Malone. 2012. "Incorporating stakeholder decision support needs into an integrated regional earth system model." *Mitigation and Adaptation for Global Change*, in press, corrected proof, DOI 10.1007/s11027-011-9345-3.

Rose SK and B Sohngen. 2011. "Global Forest Carbon Sequestration and Climate Policy Design." *Environment and Development Economics*, in press (available on-line April 14, 2011), pp 26.

Vine E. 2008. Adaptation of California's Electricity Sector to Climate Change. Public Policy Institute of California.

USEPA - U.S. Environmental Protection Agency. 2010. Supplemental EPA Analysis of the American Clean Energy and Security Act of 2009 H.R. 2454 in the 111th Congress. Accessed at: http://www.epa.gov/climatechange/economics/economicanalyses.html.

Zhai H, ES Rubin, and PL Verstee. 2011. "Water Use at Pulverized Coal Power Plants with Postcombustion." *Environmental Science and Technology* 45:2479–2485.

6.0 Research Needs

6.1 Introduction

The previous sections of this report describe at length the major complications in understanding and responding to climate-EWL interactions that are often characterized by multiple interactions, feedbacks, and tradeoffs among different human activities and environmental processes. For example, energy supply, urban development, and agricultural production often compete for land and water resources. The extent and productivity of agricultural land depends on water supply, weather, and market demand, among many other factors—and because agricultural land exchanges carbon, nutrients, energy, and water with the atmosphere and surrounding landscapes differently than other land-cover types, that balance of factors also influences the exchange of those quantities with the atmosphere and the physical climate system.

Many of the factors that determine the evolution of landscapes and their interaction with the climate depend on the demand for the quality and availability of natural resources (MEA 2005), some of which are priced in markets, and some of which are not. Therefore, accounting for these interactions at the relevant spatial and temporal scales is critical for informing and supporting effective decision making about the provision of those resources. Ultimately, this requires appropriately resolved models that are capable of capturing the processes and interactions relevant to the issue or decision being addressed, as well as observations for constraining and testing these models. In the context of climate change, this often means representing regional-scale processes effectively, and doing so in an integrated manner that accounts for human activities, including decision-making processes as well as physical, ecological, and biogeochemical processes.

Understanding the interactions and feedbacks among climate and the EWL interfaces requires not only accurate representations of each sector, but also a detailed understanding of the scale-dependent processes and interactions among them. In addition, addressing the questions that regional decision makers are asking will require the development of models capable of evaluating different adaptation strategies, testing different mitigation options, and accounting for the tradeoffs, co-benefits, and uncertainties associated with these actions or combinations of actions—such as how technology cost, accessibility to existing infrastructure, desired performance, and availability will impact results.

We have substantial knowledge, derived from both models and observations, about long-term, global constraints on changes in the climate system, on terrestrial systems, and on energy systems, and as this report has noted, we have some emerging knowledge on bilateral interfaces, such as water supply and energy infrastructure requirements. There are few analyses, however, on how the competition for multiple natural resources (such as water supply and land availability) are considered for both energy demand and production, particularly in the context of climate. For instance, are there environmental considerations, such as the availability of water or sufficient soil fertility, or economic (e.g., the availability of physical infrastructure) constraints that make the deployment of energy technologies (e.g., biofuels) or mitigation strategies (e.g., carbon capture and storage) more difficult? Such multiple interacting analyses can only be appreciated when scenarios, or multiple energy portfolios, are analyzed with greater regional specificity than the national or international strategies that are done today.

6.2 Needs and Existing Capabilities

Many studies have evaluated the current and anticipated future impacts of human-induced climate change on various sectors in different regions based on downscaled climate information. However, most of these studies do not account for processes and interactions with other sectors that could fundamentally alter the nature of the response to climate forcing. For example, many studies have projected the future impacts of climate change on agriculture in a region based on changes in global atmospheric CO_2 concentrations and regional temperature, precipitation, and other climatic factors (IPCC 2007a,b, CCSP 2008). In the absence of climate change considerations, analyses of bi-sectoral issues in energy-water date back to the late 1970s (e.g., Harte and El-Gassier 1978, references in this report). Competition for land between the energy sector and agriculture—especially related to biofuel production and water supply—is a more recent concern (e.g., Reilly and Paltsev 2009, Wise et al. 2009). Few studies, however, have accounted for other relevant factors such as changes in agricultural demand, competition over land and water resources for other uses (such as bioenergy production), and the availability and cost of new agricultural technologies. Even less appreciated are the potential confounding influences of other global changes, such as changes in dietary patterns, the availability of other food sources (e.g., the effects of overfishing, ocean acidification, and pollution on seafood production), the influence of invasive species, or changes in regional trade and agricultural incentives made for other (non-climate change related) reasons (Nelson et al. 2005). A more robust and accurate understanding of the regional impacts of climate change on agriculture—or any other sector—over the next several decades will require an integrated approach that can account for the complex, interacting processes and factors that influence outcomes and decision makers' needs on regional scales.

Many decision makers, such as natural resource managers, urban and transportation planners, and water managers, are more concerned with changes and trends that affect their region or sector in the near term—for example, over the next 10 to 30 years—than over longer time scales. The current generation of climate and integrated assessment models primarily provide insights into longer-term changes and dynamics, although the climate modeling and numerical weather forecast communities are evaluating decadal-scale predictions of changes in the physical climate system as part of the Fifth IPCC assessment report (Meehl et al. 2009). A comparable effort is needed to provide policy makers and other stakeholders with information on the near-term consequences (as well as the long-term implications) of options that might be considered to mitigate or adapt to climate change. Renewable energy portfolios, infrastructure siting decisions, and incentives for agricultural production versus biofuels are just a few examples of actions and decisions that could have implications for both adaptation and mitigation and that involve significant regional specificity. Decadal-scale simulations pose a significant challenge for integrated assessment models, which have been designed to capture long-term policy trends rather than shorter-term changes such as price shocks and business cycles.

Several recent modeling activities have begun to couple human processes with representations of climate system dynamics at global scales. Voldoire et al. (2007) coupled an integrated assessment model to the French ocean-atmosphere general circulation model to evaluate land use and land cover change interactions under a future climate change scenario that included significant changes in land use. Their results suggest that demographic and agricultural practices will dominate climate-change feedbacks on land-use decision making in the near term (less than 100 years). However, their global modeling framework could not account for regional considerations. Similarly, there has been extensive work on harmonizing global land-use history reconstructions, as part of the Representative Concentration Pathway

(RCP) process (Moss et al. 2010) for use in the CMIP5 simulations (Hurtt et al. 2011; Thomson et al. 2011). These data sets are global, and were compiled or interpolated to a 0.5 degree grid for use in global climate models. Literature is now emerging that examines the interactions of land-use and energy decisions, with broad implications for the total amount of forested and agricultural land across the globe (e.g., Wise et al. 2009), or for the future of regional landscapes (Thomson et al. 2011).

6.3 Needs for Integrative Portfolio of Options

Regional analyses that consider integrated assessment and climate that are geographically specific have also been documented (e.g., Holman et al. 2005*a,b*; Izaurralde et al. 2003). There are, however, no integrative tools that are portable (that is, do not require extensive calibration) to any region; that address the dynamic interactions between the energy, land, water, climate and socio-economic systems; and that allow for an integrated evaluation of the processes and outcomes that decision makers are asking for. Meaningful analyses of these and other regional issues will require a new class of models, measurements, and observations that are consistent with global climate and socio-economic constraints, and capable of resolving regional human decision making and natural processes in a manner that captures the full range of relevant interactions and feedbacks (e.g., see Figure 2.1).

Designing a framework represented by Figure 2.1 adequately describes and simulates the interactions of climate, ecosystem, energy, and economy at regional scales is a major challenge. One rationale for doing so is also to provide informed scientific input to regional decision making, especially about adaptation and mitigation issues. It would be difficult to capture the insight needed without meaningful engagements with stakeholder and decision-making communities. Experience with the needs and requirements of these groups will provide rapid insight into which processes in the models must be improved or added, as well as insight into managing expectations of integrated outcomes. In the first U.S. National Assessment of the Impacts of Climate Change and Variability, for example, a substantial effort was made to identify issues that regional stakeholders around the country were concerned about (NAST 2000, Morgan et al. 2005). Many stakeholders around the country identified land-use and water management issues as being particularly importance, and in most cases as being more important than just understanding the potential for the impacts of climate change alone. But these are environmental issues that require a consideration not only of variability in the climate system, but also an understanding of how supply, demand, and the availability of energy technologies interact with each other over time.

The experimentation and evaluation strategies that emerge to address climate-EWL interactions will necessitate new thinking in risk, uncertainty and vulnerability analyses (see section 4). Whereas uncertainty quantification that identifies error propagation in data and optimal parameter estimation in models is clearly needed, this need will be compounded by the need to understand both the quantitative and qualitative uncertainty introduced by the criteria that decision makers actually use (e.g., price and distance of existing infrastructure for new energy development), by the risks they are willing to take, and by their ability to explore alternative scenarios of the future. Therefore, a more complete characterization of uncertainty will not only encompass an understanding of how the model behaves with different parameters and data sets, but also an understanding of the types of decision-making to which it is being applied.

What are some of the highest priority challenges involved in developing and testing such a modeling framework? Such an effort, after all, will require considerable effort from and coordination across the natural and social sciences, as well as the software and energy engineering communities. We identify three areas as providing some of the most serious challenges to such an interdisciplinary effort: mismatched time and space scales of model components, mismatched communications among disciplines, and the availability of data.

The native temporal and spatial scales of natural and human systems vary tremendously. For instance, dynamically downscaled regional models of the climate system typically resolve topography and explicitly simulate climate processes at spatial scales down to 10 kilometer regularly spaced grids using half-minute (e.g., 30 sec) time steps (e.g., see Skamarock et al. 2008). The current generation of integrated assessment models, in contrast, are global in scope but resolve the energy economies of a relatively small number (14–22) of geopolitical regions (see Vanvuuren et al. 2011 and references therein). Other key systems span these spatial and temporal scales. For example, agriculture, land use, and precipitation/runoff can be considered from the individual farm, ecosystem, or sub-basin level, respectively, but are more commonly represented on scales of tens to hundreds of kilometers in climate models. Energy operations models operate at spatial scales from individual nodes to utility zone and temporal scales from hourly annual time steps. Finer scale energy-economy models can be Excel based (e.g., Hoffman et al., 2010) and are proprietary whereas coarser models such as the National Energy Modeling System (NEMS), developed by the U.S. Energy Information Administration (EIA), provide long-term national and regional projections of production, consumption, and prices for many energy products, as well as emissions, GDP, unemployment, imports, exports, and other outputs (e.g., EIA 2009).

Even when experts in different disciplines are willing to collaborate and collegial relationships develop, there are still many practical, philosophical, and cultural differences to overcome. Integration across different fields of science, scholarly research, and engineering raises a number of practical issues, such as differences in vocabulary and nomenclature. For example, the "reference" or "control" run of a climate model typically refers to a simulation in which the influence of human activities is neglected or artificially suppressed. For integrated assessment models, in contrast, a "reference case" typically refers to a simulation of how future human activities and GHG emissions will evolve in the absence of any explicit climate change mitigation policies. The climate and integrated assessment modeling communities also typically use different terms to refer to the assumptions and external (or unresolved) processes in their models—many of the "exogenous factors" in an economic model are what a climate modeler would call "external forcings." Regular communication and explicit agreement on definitions and assumptions are thus critical for success.

Another challenge will be developing a strategy, metrics, and appropriate datasets for evaluating integrated model performance. Lessons learned from coupling different components of global Earth system models (i.e., atmosphere, ocean, land, and sea ice) will be useful, but novel approaches will be needed. Physical or biogeochemical modeling groups typically test their models by initializing with historical climate or ecological information and comparing results to observations. Human system modelers, in contrast, often do not have the requisite data, or model structure to evaluate their models in this manner. Their models have used the relatively sparse historical record for initialization, and there has not been a concerted community effort to develop independent data sets for model evaluation. Integrated assessment model intercomparisons of similar experiments have attempted to serve this purpose (e.g., Clarke et al. 2009). Accounting for different types and modes of economic variability is a related complication, one that is only somewhat analogous to the differences between validating a long-term climate simulation versus a short-term weather forecast. Of course, one of the lessons from the AR4 future climate simulations is that the ability of climate models to reproduce the past accurately does not guarantee the validity of projections of future climate change (e.g., Friedlingstein et al. 2006).

Even given the formidable challenges of communication and development, the biggest limiting factor for an integrated regional approach may be data—specifically, the quality, resolution, and sheer amount of data needed for model development, initialization, and evaluation. Data transformations and software architecture are needed to represent data transformations and feedbacks among different model components across such disparate time and spatial scales. Meeting this challenge will require communication and collaboration across disciplines, particularly in the software engineering communities. For new model components, it will likely be necessary to consider novel approaches toward interpolation or extrapolation of extant data sets. Using land use as a benchmark for model evaluation is another possible avenue to move forward—there exist extensive and intensive land cover data and information from many sources, and comparing model performance to these datasets of current and past land use changes can help to develop confidence in projected future land use trajectories.

6.4 Summary

In summary, a case study described in section 3.2 of this report illustrates the current and ongoing conflicts that local to statewide communities are struggling to address in Texas with regard to energy production, water supply, and demand for land resources in the context of a current drought and socio-economic changes (e.g., demographic, economics). While drought policies and water planning policies are in place (e.g., National Drought Policy Commission 2004, Texas Water Development Board 2012), there is no plan or policy that can provide a family of solutions or scenarios for possible economic and natural resource options in the future that is scalable across local to regional domains with state or national legislative or environmental constraints. Using the Texas case study to represent future research needs, the following priorities are recommended:

- Meaningful analyses of the EWL interfaces will require a new class of models, measurements, and observations that are consistent with global climate and socio-economic constraints, and capable of resolving regional human decision-making and natural processes in a manner that captures the full range of relevant interactions and feedbacks. This can be initialized through adaptation of existing modeling and observing frameworks.

- New models, observing systems, or, even modifications of existing frameworks will require new strategies for understanding and quantifying uncertainty.

- Evaluation strategies that account not only for individual model performance, but properties of coupled systems will require robust metrics for benchmarking model performance and data systems.

- For regions, or areas that are unable to provide the current data required (e.g., recent transportation infrastructure, pricing policies, demography, environmental information), new methods into parsing sparse data or extracting information from pre-existing data (e.g., extrapolating new information from old data) will be required.

- In the context of climate, support for new research and modeling capabilities that accounts for potential future environmental constraints (e.g., availability of water), economic limitations (e.g., existing infrastructure) and scenario development will be needed to inform decision making processes for the deployment of future energy transitions.

- New tools that provide stakeholders with information on near-term consequences, as well as long-term implications of options that might be considered to mitigate or adapt to climate change will be needed. This will require engagement by the research communities with decision makers and policy communities.

- Finally, accounting for natural boundaries, such as watershed, energy utility or geo-political zones will need to be incorporated into existing gridded calculations and observations. This can only be accomplished through significant interactions and engagement across the natural and social sciences, engineering, humanities (e.g., economics) and stakeholder communities.

6.5 References

CCSP – U.S. Climate Change Science Program. 2008. The effects of climate change on agriculture, land resources, water resources, and biodiversity in the United States, A Report by the U.S. Climate Change Science Program and the Subcommittee on Global Change Research. P Backlund, A Janetos, D Schimel, J Hatfield, K Boote, P Fay, L Hahn, C Izaurralde, BA Kimball, T Mader, J Morgan, D Ort, W Polley, A Thomson, D Wolfe, MG Ryan, SR Archer, R Birdsey, C Dahm, L Heath, J Hicke, D Hollinger, T Huxman, G Okin, R Oren, J Randerson, W Schlesinger, D Lettenmaier, D Major, L Poff, S Running, L Hansen, D Inouye, BP Kelly, L Meyerson, B Peterson, R Shaw. U.S. Department of Agriculture, Washington, DC, 362 pp.

Clarke L, J Edmonds, V Krey, R Richels, S Rose, and T Massimo. 2009. "International climate policy architectures: Overview of the EMF 22 International Scenarios." *Energy Economics* 31:S64-S81, DOI:10.1016/j.eneco.2009.10.013.

EIA - Energy Information Administration. 2009. Integrating Module of the National Energy Modeling System: Model Documentation. DOE/EIA-M057, Washington, DC, May 2009.

Friedlingstein P, P Cox, R Betts, L Bopp, W von Bloh, V Brovkin, P Cadule, S Doney, M Eby, I Fung, B Govindasamy, J John, C Jones, F Joos, T Kato, M Kawamiya, W Knorr, K Lindsay, HD Matthews, T Raddatz, P Rayner, C Reick, E Roeckner, K-G Schnitzler, R Schnur, K Strassmann, AJ Weaver, C Yoshikawa, and N Zeng. 2006. "Climate -carbon cycle feedback analysis, results from the C4MIP model intercomparisons." *Journal of Climate* 19:3337-3353.

Harte J and M El-Gasseir. 1978. "Energy and Water." *Science* 199(10):623-634.

Hoffman, MG, MC Kintner-Meyer, A Sadovsky, and JD DeSteese. 2010. Analysis tools for sizing and placement of energy storage in grid applications: A literature review. PNNL Technical Report PNNL-19703. 32 pp.

Holman IP, MDA Rounsevell, S Shackley, PA Harrison, RJ Nicholls, PM Berry, and E Audsley. 2005a. "A regional, multi-sectoral and integrated assessment of the impacts of climate and socio-economic change in the UK: Part I: Methodology." *Climatic Change* 71:9-41, DOI: 10.1007/s10584-005-5927-y.

Holman IP, RJ Nicholls, PM Berry, PA Harrison, E Audsley, S Shackley, and MDA Rounsevell. 2005b. "A regional, multi-sectoral and integrated assessment of the impacts of climate and socio-economic change in the UK: Part II: Results." *Climatic Change* 71:43-74, DOI: 10.1007/s10584-005-5956-6.

IPCC – Intergovernmental Panel on Climate Change. 2007a. Summary for Policymakers. In: Climate Change 2007: Impacts, Adaptation and Vulnerability. Contribution of Working Group II to the Fourth Assessment Report of the Intergovernmental Panel on Climate Change, eds. ML Parry, OF Canziani, JP Palutikof, PJ van der Linden and CE Hanson. Cambridge University Press, Cambridge, UK, 7-22.

IPCC – Intergovernmental Panel on Climate Change. 2007b. Climate Change 2007: Synthesis Report. Contribution of Working Groups I, II and III to the Fourth Assessment Report of the Intergovernmental Panel on Climate Change [eds. RK Pachauri and A Reisinger, Core Writing Team]. IPCC, Geneva, Switzerland, 104 pp.

Izaurralde RC, NJ Rosenberg, RA Brown, and AM Thomson. 2003. "Integrated assessment of Hadley Center (HadCM2) climate-change impacts on agricultural productivity and irrigation water supply in the conterminous United States. Part II. Regional agricultural production in 2030 and 2095." *Agricultural and Forest Meteorology* 117:97-122, DOI:10.1016/S0168-1923(03)00024-8.

MEA - Millennium Ecosystem Assessment. 2005. Ecosystems and human well-being: Scenarios, Vol. 2, 173–222. Island Press, Washington, DC.

Meehl GA, L Goddard, J Murphy, RJ Stouffer, G Boer, G Danabasoglu, K Dixon, MA Giorgetta, AM Greene, E Hawkins, G Hegerl, D Karoly, N Keenlyside, M Kimoto, B Kirtman, A Navarra, R Pulwarty, D Smith, D Stammer, and T Stockdale. 2009. "Decadal prediction: Can it be skillful?" BAMS, DOI:10.1175/2009BAMS2778.1

Morgan MG, R Cantor, WC Clark, A Fisher, HD Jacoby, AC Janetos, AP Kinzig, J Melillo, RB Street, and TJ Wilbanks. 2005. "Learning from the U.S. National Assessment of Climate Change." *Environmental Science & Technology* 39:9023-9032.

NAST - National Assessment Synthesis Team. 2000. Climate Change Impacts on the United States: The Potential Consequences of Climate Variability and Change, Overview Report. US Global Change Research Program, Washington, DC.

Nelson GC, E Bennett, AA Berhe, KG Cassman, R DeFries, T Dietz, A Dobson, et al. 2005. "Drivers of change in ecosystem condition and services." In *Millennium.*

Reilly J and Paltsev. 2009. "Biomass energy and competition for land." In *Economic Analysis of Land Use in Global Climate Change Policy*, eds. T Hertel, S Rose, R Tol. Routledge, UK, pp. 184-207.

Skamarock WC, JB Klemp, J Dudhia, DO Gill, DM Barker, MG Duda, X-Y Huang, W Want, and JG Powers. 2008. A description of the Advanced Research WRF Version 3. NCAR Technical Note 475+STR. 113. pp.

Van Vuuren DP, J Edmonds, M Kainuma, K Riahi, A Thomson, K Hibbard, G Hurtt, T Kram, V Krey, J-F Lamarque, T Masui, M Meinshausen, N Nakicenovic, SJ Smith, and SK Rose. "The representative concentration pathways: an overview." *Climatic Change* 109:5-31.

Voldoire A, B Eickhout, M Schaeffer, J-F Royer, and F Chauvin. 2007. "Climate simulation of the twenty-first century with interactive land-use changes." *Climate Dynamics* 29(2-3):177-193, DOI: 10.1007/s00382-007-028-y.

7.0 Key Findings

Characterization of Climate and Energy-Water-Land System Interactions

- Population growth and economic and social development are major drivers of the demand for energy, land, and water resources within the interdependent climate and EWL system. Allocation of limited resources among competing uses will require tradeoffs with climate variability and change implications depending on location and time frame. A major challenge will be to manage, and optimize where possible, competing economic and environmental objectives and priorities within resource budget constraints and impact risks of climate variability and change.

- The interdependencies of climate and the EWL system can be characterized by the three bilateral interfaces of energy-water, energy-land, and land-water. Each bilateral interface consists of linkages representing the supplies, end-use demands, and associated functional relationships between the two sectors. The linkages can be defined in terms of resource *demand*, supply *endowment*, and *technologies* (including mechanisms, processes, and systems). Consideration of climate interactions with only one or two of the bilateral interfaces in isolation provides only a partial impact assessment.

- A comprehensive assessment requires consideration of the three bilateral interfaces into an integrated EWL system that includes the interdependencies and feedbacks among all three resource sectors and climate.

- Much of our current understanding of climate impacts on the complex interdependencies of the EWL system are derived from limited observations of bilateral interface responses to climate variability. The concept of EWL interfaces can help identify the relative degree of risks and vulnerabilities to the effects of climate variability and change. It can also potentially help identify opportunities for mitigating or managing climate change impacts through technical and policy interventions.

Energy-Water-Land Interfaces: Resource Interdependencies and Interactions with Climate

- Focusing on sector to sector interfaces alone does not adequately capture the complexity and importance of the EWL system. The many bilateral interfaces form a dynamic set of interacting processes linked through a complex network of feedbacks.

- Competition for water is the most straightforward conflict linking energy, water, and land (e.g., simultaneous demand for thermoelectric generation, irrigation, environmental flows).

- Extreme climate events such as drought and associated heat waves have important impacts on the EWL interfaces. Impacts are seen as changes in cropping and grazing and accompanying wildfire damage. These changes tend to reinforce and intensify individual impacts on land and water resources (e.g., reduced cropping raises feed prices, which changes grazing patterns, which in turn affects vegetation density and thus wild fire vulnerability). To a lesser extent, these changes feed back through water and land use to impact energy demand and production.

- U.S. regions differ in their (a) current climate, (b) projected climate change, (c) energy mix (e.g., solar/wind availability, coal), (d) energy supply and demand, (e) water availability/regularity and water sources (e.g., rain vs. snow fed), and (f) the availability and quality of land. Each region will be differentially impacted by climate change and each region will have to adapt or mitigate using different strategies. The manner in which adaptation strategies and associated institutions evolve has significant implications for energy-water-land dynamics.

- California and the Gulf states would likely follow very different compliance paths if stringent emission standards were adopted. Because of abundant renewable resources and past proactive adoption on the part of the state, California would expand electricity production with wind, solar, geothermal, biomass, and small hydroelectric. In contrast, limited renewables in the Gulf states may lead to broad implementation of CCS utilizing abundant deep saline aquifers for storage. In terms of transportation, California would likely turn to plug-in hybrid electric vehicles and all-electric vehicles to meet emission standards because of limited water resources. Gulf states would favor cultivation of biomass for biofuel production making use of their relative abundance of water.

Risk, Uncertainty, and Vulnerability Associated with Climate Impacts on Energy-Water-Land Interfaces.

- Risk, uncertainty, and vulnerability at the EWL interfaces are minimally reported in the literature, and where they are documented, they are usually case-specific. However, risk, uncertainty, and vulnerability are generally found to have the following four characteristics:

  – They are broader in scope.

  – They can be amplified or attenuated across sectors.

  – They have altered temporal and spatial dynamics.

  – They manifest during extreme (low-likelihood, high-consequence) events.

  These characteristics are fundamental to understanding how risk, uncertainty, and vulnerability relate to each characteristic across sectors, and for developing solutions and strategies that may reduce their impact or influence.

- Examples from the literature help to illustrate how the relationships among the stochastic (e.g., the duration and magnitude of the drought), epistemic (e.g., what are the future energy and water requirements), and human response uncertainties (e.g., to what degree are the water and energy conservation policies adopted) are cumulative functions of one another that make it more difficult to answer questions about the impacts of climate and the impacts of the solutions. In other words, risk, uncertainty, and vulnerability across sector interfaces are broader in scope than single sector estimates of the same.

- Amplification and attenuation refer to how risk, uncertainty, and/or vulnerability for systems in one sector react to changes in a different sector. The relationships between the uncertainty in one sector and the risk and vulnerabilities in other sectors are highly non-linear and difficult to predict.

- The nature of cross-sector systems are such that the associated risks, uncertainties, and vulnerabilities span large scales of time and space that create interdependencies and secondary dynamics that otherwise would not exist.

- Risk, uncertainty, and vulnerability across sectors are dominated by low-probability, high-consequence events, mainly due to the built-in resilience of existing technology and infrastructure to higher probability, lower consequence events.

- Awareness of the four characteristics of risk, uncertainty and vulnerability can enable monitoring and assessment mechanisms to anticipate and avoid or correct unintended consequences, and that more effective emergency planning and response can be put in place by understanding the sources of those consequences.

- Perhaps the biggest gains will come by increasing our understanding of human response and behavior to climate change and decisions concerning climate change, identifying the trigger points where low-probability events within one sector can become high-consequence events in other sectors, and in identifying and understanding the amplification, attenuation, and feedback mechanisms that create unintended and unanticipated consequences.

Climate Mitigation and Adaptation at the Energy-Water-Land Interfaces

- Many mitigation and adaptation options tie directly into one of the EWL sectors, and are therefore tied into the EWL interfaces. Understanding the EWL nexus is therefore central to the effective design, selection, implementation, and monitoring of adaptation and mitigation. Almost all mitigation options lie within either the energy or land sectors. Mitigation reduces or sequesters emissions arising from the supply and demand for energy and land (e.g., substituting renewable technologies for fossil fuel generation; preventing deforestation). As such, mitigation options are affected by EWL relationships.

- Adaptation options designed to reduce vulnerability to climate impacts in one EWL sector affect, and are affected by, EWL linkages. Some adaptation measures reduce demands on EWL endowments (e.g., water-use efficiency), while others may increase them (e.g., desalinization).

- Understanding mitigation and adaptation relationships to the EWL interfaces facilitates not only the evaluation of the net impact of individual mitigation or adaptation measures, but also the compound effects of concurrent implementation, either intentionally or as an outcome of the uncoordinated actions of independent parties. These compound effects may not always have positive synergies, and ignoring or failing to identify potentially negative interactions runs the risk of undermining the original policy goals

- Some sector-specific mitigation and adaptation measures have the potential to provide synergistic "win-win" opportunities to enhance climate mitigation or adaptation objectives across one or more other sectors in the nexus. Other measures may have negative impacts on mitigation or adaptation potential in other sectors. Such cross-sectoral impacts can carry substantial risks of inadvertently diminishing climate mitigation and/or adaptation objectives.

Research Needs Associated with Climate Impacts on Energy-Water-Land Interfaces

- A major complication in understanding and responding to climate changes is that they are often characterized by multiple interactions, feedbacks, and tradeoffs among different human activities and environmental processes.

- Simulating and understanding the interactions and feedbacks among climate and the EWL system requires not only accurate representations of each individual sector, but also a detailed understanding of the scale-dependent interactions among them.

- Addressing the climate-EWL related questions that regional decision makers are asking will require the development of models capable of evaluating different adaptation strategies, testing different mitigation options, and accounting for the tradeoffs, co-benefits, and uncertainties associated with these actions or combinations of actions—such as how technology cost, performance, and availability will impact results.

- We have substantial knowledge, derived from both models and observations, about long-term, global constraints on changes in the climate system, on terrestrial systems, and on energy systems, and we have some emerging knowledge on bilateral interfaces, such as water supply and energy infrastructure requirements. There are few analyses, however, on how the competition for multiple natural resources (such as water supply, land availability) is considered for both energy demand and production.

- The current generation of global climate and integrated assessment models do not adequately address regional issues such as regional options for energy portfolios, resolving conflicts associated with land and water resources, evaluating the tradeoffs and synergies associated with different adaptation and mitigation strategies, and projecting the impacts of climate change on water, ecosystems, human health, and other human and environmental systems.

- Research needed to substantially increase our understanding of the interactions and feedbacks among energy, water, land, and climate include the following:

  - Meaningful analyses of the EWL interfaces will require a new class of models, measurements, and observations that are consistent with global climate and socio-economic constraints, and capable of resolving regional human decision-making and natural processes in a manner that captures the full range of relevant interactions and feedbacks.

  - New models, observing systems, or, even modifications of existing frameworks will require new strategies for understanding and quantifying uncertainty.

  - Evaluation strategies that account not only for individual model performance, but properties of coupled systems will require robust metrics for benchmarking model performance and data systems.

  - For regions, or areas that are unable to provide the current data required (e.g., recent transportation infrastructure, pricing policies, demography, environmental information), new methods into parsing sparse data or extracting information from pre-existing data will be required.

  - In the context of climate, support for new research and modeling capabilities that account for potential future environmental constraints (e.g., availability of water), economic limitations (e.g., existing infrastructure) and scenario development will be needed to inform decision making processes for the deployment of future energy transitions.

  - New tools that provide stakeholders with information on near-term consequences, as well as long-term implications of options that might be considered to mitigate or adapt to climate change will be needed.

  - Accounting for natural boundaries, such as watershed, energy utility or geo-political zones will need to be incorporated into existing gridded calculations and observations.

Appendix A

Demand- Endowment-Technology (DET)
Interface Linkage Model

Appendix A
Demand-Endowment-Technology (DET)
Interface Linkage Model

The DET representation illustrated in Figure A.1 includes resource supply attributes, end-use requirements, and associated functional relationships and dependencies from supply to end-use. Examples of resource supply attributes include the types, locations, availability, and accessibility of sources and the quantity and quality of stocks and flows. Other supply characteristics include whether the source stocks and flows are renewable, finite and depletable, substitutable, or otherwise constrained in terms of withdrawal rates, timing and duration of withdrawals, and costs. Demand attributes include the specific types and locations of end-use applications, end-use requirements and constraints (compatibility with infrastructure, costs, etc.), quantity and quality of stocks and flows needed, and time frames (use rate and duration profiles). On both the supply and the demand side of the interface will be the natural and human-mediated technologies, processes, systems, and infrastructures required to produce, supply, and deliver the resource to meet the end-use demand.

The complex, multi-dimensional functional dependency between resource supplies (S) and resource demands (D) is represented notionally in Figure A.1 by the function $f_{SD}(N_S, N_D, E_S, E_D)$. This functional notation takes into account the natural systems and processes on both the supply (N_S) and the demand (N_D) side of the interface, along with the human engineered technologies and processes on both the supply (E_S) and the demand (E_D) side. The functional elements representing both the natural and the engineered technologies on each side of the interface are combined to give the shortened functional linkage notation $f_{SD}(N_{SD}, E_{SD})$. The functional element N_{SD} notionally represents the combination of all of the *natural* processes and systems that do not directly involve human actions, infrastructure, or technology interventions, while E_{SD} represents all of the *engineered* technologies and processes that involve human intervention. Engineered technology includes human-made infrastructure, physical and chemical production and conversion processes, and other systems that represent technical productivity. It also includes the economic, legal, and policy structures, mechanisms, and substitutability in production that represent economic productivity. Table A.1 provides a summarized list and high-level description of the twelve combined natural and human engineered technology and functional process elements that represent all six bilateral supply-demand cross-linkages for the overall integrated climate-EWL nexus system illustrated in Figure 2.3.

All of the interface linkage relationships, as well as the resource supply and demand stocks and flows within each sector and between sectors, will be assumed to involve some degree of interdependency with climate, as illustrated by Figure 2.3. Some interface linkages will presumably have stronger interactions with, and be more impacted by, climate variability and climate change than others. This will depend on the specific scenario being considered, the key natural and engineered processes that predominantly come into play, the relative influence of climate variability and change on these processes, and the spatial and temporal scales involved.

The more that the details of interface linkages can be understood and expressed analytically or modeled numerically, and the more that they can be further linked directly to the effects of climate variability and change, the more quantitative the analysis can be. To the extent that this can be done, the concept of functional linkages between the EWL nexus interfaces may help identify the specific points of interaction coupling having the greatest risks and vulnerabilities to the effects of climate variability and change. It may also help identify opportunities for mitigating or managing climate change impacts through interventions that may include technical, policy, and changing resource use pattern.

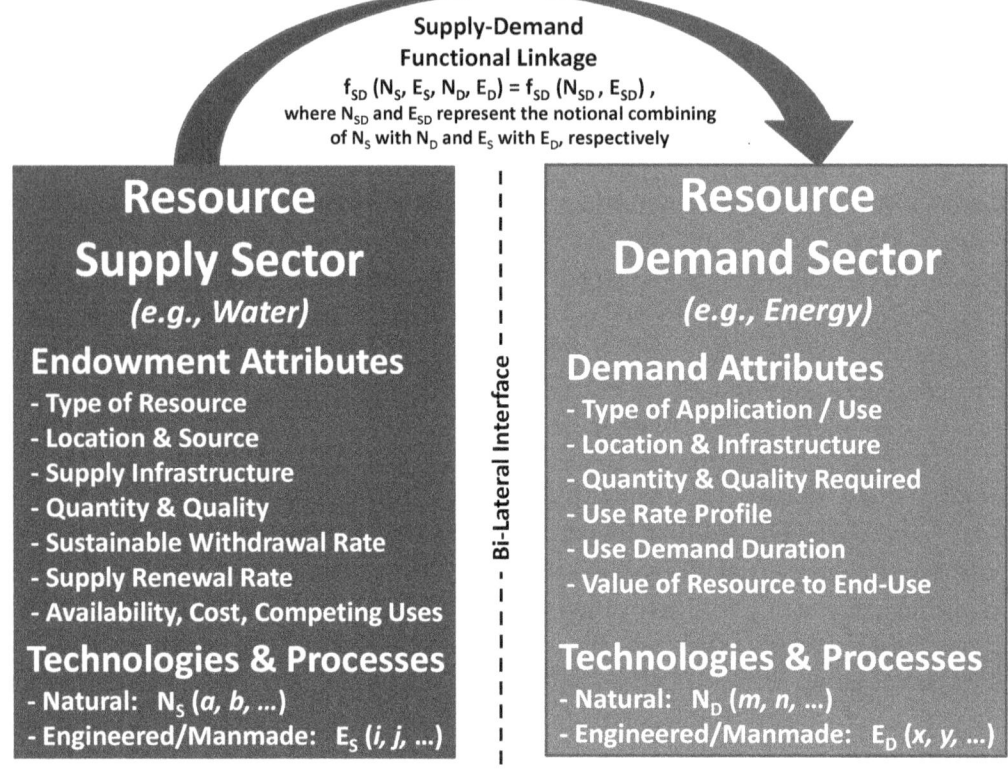

Figure A.1. Notional illustration of a bilateral resource sector interface linkage structure and functional relationship characterized as the *demand-endowment-technology* (DET) model.

Table A.1. Bilateral Interface Supply (S)-Demand (D) Functional Linkage Elements

Functional Nomenclature for Natural and Human Engineered Technologies and Processes	
Symbol	N_{SD} and E_{SD} Functional Descriptions (refer to Appendix A and Figure 2.3)
N_{WE}	**Natural processes and technologies for water resources to meet energy demands** Examples: weather and hydrologic cycle mechanisms and processes of precipitation, watershed capture/transfer to replenish and maintain surface and ground water resources (including snowpack storage and timed release), soil moisture flows/latencies, evapo-transporation to atmosphere.
E_{WE}	**Engineered processes and technologies for water resources to meet energy demands** Examples: Dams, reservoirs, hydropower generation facilities; groundwater wells, pumps and pipelines; surface water control and transport infrastructure; water systems and infrastructure for power plant cooling, energy mineral extraction, and irrigation for biomass production for energy.
N_{EW}	**Natural processes and technologies for energy resources to meet water demands** Examples: groundwater aquifer source accumulation, filtering, and recharge mechanisms and processes to support pumped and transported groundwater supplies; surface water source flows and renewal mechanisms and processes to support pumped and transported surface water supplies; natural processes and mechanisms associated with filtering, cleaning, desalination, deionization, disinfection, heating, and cooling of water supplies; natural processes associated with production of energy and fuels for electrical and mechanical power, heating, and cooling for pumping,

Functional Nomenclature for Natural and Human Engineered Technologies and Processes	
Symbol	**N_{SD} and E_{SD} Functional Descriptions (refer to Appendix A and Figure 2.3)**
	transporting, processing water.
E_{EW}	**Engineered processes and technologies for energy resources to meet water demands** Examples: technologies, processes and systems for electrical and mechanical power and energy production and delivery for water pumping, transport, processing, and conditioning; engineered technologies, processes, and systems for water treatment (filtering, cleaning, desalination, disinfection) and conditioning (heating, cooling, deionization).
N_{LE}	**Natural processes and technologies for land resources to meet energy demands** Examples: Land required for siting nuclear and thermoelectric power plants, oil refineries, biorefineries, wind farms, and solar arrays. Right-of-way for constructing pipelines, high-voltage transmission lines.
E_{LE}	**Engineered processes and technologies for land resources to meet energy demands** Examples: mines (coal, uranium, oil shale, tar sands); wells (petroleum, conventional gas, shale gas, coal bed methane); sites for refineries and biorefineries, bioenergy crop production, harvesting, and processing infrastructure; power plants, dams and reservoir sites for hydropower production, sites for wind farms and solar power plants ; transmission and distribution lines, charging stations for electric vehicles; fuel processing, transport, storage, and distribution infrastructure (pipelines, tanks, railways, barges, roadways, tanker trucks);
N_{EL}	**Natural processes and technologies for energy resources to meet land demands** Examples: Wind (air movement), solar (photon flux, atmospheric processes for transmission & scattering), geothermal energy (subsurface thermal energy: hot rocks, hot water, steam), flowing water (kinetic energy) and elevated/falling water (potential/kinetic energy); combustion and heat flow processes;
E_{EL}	**Engineered processes and technologies for energy resources to meet land demands** Examples: Electric power generation (hydropower, fossil and nuclear thermoelectric), transmission, and distribution infrastructure to supply farms, ranches, rural communities, towns, cities, urban centers; Liquid transportation fuel production, processing, transmission, storage and distribution; Transportation vehicles (air, ground, marine) and engines/motors that utilize liquid and gaseous fuels and electricity; Solar, wind, geothermal
N_{WL}	**Natural processes and technologies for water resources to meet land demands** Examples: Water storage in snowpacks, lakes, streamflow, and soil to meet demands from land (irrigated and dryland agriculture, livestock, domestic, commercial, industrial)
E_{WL}	**Engineered processes and technologies for water resources to meet land demands** Examples: Groundwater extraction to meet demands from land (irrigated and dryland agriculture, livestock, domestic, commercial, industrial).
N_{LW}	**Natural processes and technologies for land resources to meet water demands** Examples: Watershed processes for capture in lakes, streams, rivers and other surface flows, percolation and subsurface recharge of groundwater, storage in snow pack with slow release over time. Underground aquifers fed by recharge pathways.
E_{LW}	**Engineered processes and technologies for land resources to meet water demands** Examples: Dams, reservoirs, water transport (diaducts, large canals) and flow control infrastructure, pipelines, pumps, wells, irrigation systems (irrigation channels, subsurface, center pivot irrigation, drip irrigation systems).

Appendix B

High-Level Inventory Matrix for Energy-Water-Land

Appendix B
High-Level Inventory Matrix for Energy-Water-Land

Table B.1. High-Level Inventory Matrix for Energy-Water-Land Represented as Three Bilateral, Bidirectional Interfaces

Applications/Systems/Processes	Bi-Directional, Bilateral Interdependencies [Examples of Climate Linkage Shown in Brackets in Table]		
	Energy-Water (E↔W)	Energy-Land (E↔L)	Land-Water (L↔W)
	Interface Linkage Functions f_{EW} (N_{EW}, E_{EW}) and f_{WE} (N_{WE}, E_{WE}) [see Fig 2.3, Fig A.1, Table A.1]	Interface Linkage Functions f_{EL} (N_{EL}, E_{EL}) and f_{LE} (N_{LE}, E_{LE}) [see Fig 2.3, Fig A.1, Table A.1]	Interface Linkage Functions F_{LW} (N_{LW}, E_{LW}) and f_{WL} (N_{WL}, E_{WL}) [see Fig 2.3, Fig B.1, Table A.1]
Electric Power Generation **Hydroelectric** **Thermoelectric (Fossil, Nuclear, & Biopower), Geothermal, Renewable Solar & Wind** **Transportation (Land, Air, Marine)** **Petroleum fuels** **Biofuels** **Electric & hybrid vehicles** **Municipal, Industrial, Commercial, Residential, and Government** **(Federal/State/Local) Facilities and Operations** **Water Resources Pumping, Treatment, Use Conditioning** **Carbon Capture & Storage (CCS)**	• Energy/power for water pumping, transport, treatment and conditioning for end-use [extreme weather event impacts on infrastructure] • Water for hydropower [drought – reduced supplies; higher temps – reduced snowpack and earlier snowmelt altering supply flow timing; increased reservoir evaporation loss] • Water for thermal power cooling [fossil, nuclear, biomass]; [drought – reduced supplies; higher temps – reduced snowpack and earlier snowmelt altering supply flow timing; heat waves – increased power demand & stress on water supplies] • Water for geothermal power • Water for solar thermal generation [drought – reduced supplies; higher temps – reduced snowpack and earlier snowmelt altering supply flow	• Energy/power/CHP for communities, towns, cities, municipalities, industrial and commercial facilities, residential • Power and fuels generation for conventional, electric and hybrid vehicles and mass transit • Land use for producing biomass for biopower [drought, heat wave, flooding impacts on biomass crop production] • Land for thermal power plants [high temperatures reduce thermal cooling efficiency;] • Land for hydropower [drought in watershed; reduced watershed precipitation & snow pack; flooding & siltation] • Land for geothermal power • Land for transmission and distribution lines • Land for roadways, railways, airports, fueling stations	• Land for watershed to capture precipitation/ snow pack and provide surface water flows and groundwater recharge [drought in watershed; reduced watershed precipitation & snow pack; increased evapotranspiration] Land for surface water storage reservoirs, lakes, streams, rivers and ground water [prolonged drought ; overpumping] • Ecological interactions with cooling water discharge • Ecological and agricultural interactions with dam release and/or snow pack melt timing and flow variations [drought – reduced supplies; higher temps – reduced snowpack and earlier snowmelt altering supply flow timing; flooding; nutrient-loaded tail water from crop

	Bi-Directional, Bilateral Interdependencies [Examples of Climate Linkage Shown in Brackets in Table]		
Applications/Systems/Processes	Energy-Water (E↔W) Interface Linkage Functions f_{EW} (N_{EW}, E_{EW}) and f_{WE} (N_{WE}, E_{WE}) *[see Fig 2.3, Fig A.1, Table A.1]*	Energy-Land (E↔L) Interface Linkage Functions f_{EL} (N_{EL}, E_{EL}) and f_{LE} (N_{LE}, E_{LE}) *[see Fig 2.3, Fig A.1, Table A.1]*	Land-Water (L↔W) Interface Linkage Functions F_{LW} (N_{LW}, E_{LW}, E_{WL}) and f_{WL} (N_{WL}, E_{WL}) *[see Fig 2.3, Fig B.1, Table A.1]*
	timing; heat waves – increased power demand & stress on water supplies] • Reduced water demand from expansion of wind and PV [Reduction in PV performance at elevated temperatures; Impacts of changing insolation and wind conditions; infrastructure damage from extreme weather events] • Water for biomass for biopower and transportation biofuels [drought, heat, and flood impacts on biomass production; extreme weather event impacts on transportation fuel production infrastructure; reductions in water supplies – increased water competition for shale gas fracking, etc.] •	• Energy/fuels for operation of ground and air vehicles and transportation support infrastructure [extreme weather events – damage to infrastructure] • Land for wind and solar energy [wind and solar resource changes impacts wind & solar generation] • Land for nuclear waste repositories • Land and infrastructure for fossil power plant CCS [extreme weather events – damage to infrastructure]	irrigation]
Agriculture, Aquaculture, Forestry Eco-System Health and Services	• Water for biofuels/bioenergy feedstock production and conversion processing [drought, prolonged heat wave; leading to crop damage] • Feedback impact on ecosystems of wastewaters from bioenergy production (nutrient loading), produced and fracking water from energy mineral extraction (contamination), and water used for power plant cooling (thermal loading) [algae blooms in nutrient rich water during periods of high temperatures; spillage of energy mineral extraction	• Energy/power for agriculture, aquaculture, forestry operations Atmospheric carbon capture and storage (CCS) in land cover vegetation canopy and roots/soil [extreme weather event impacts; flooding/erosion; drought & heat leading to water shortages and plant kill] • Use of agriculture land for siting of renewable energy production (wind, solar, biomass) [extreme weather event impacts on infrastructure; flooding; high	• Land for agriculture (farm, ranch, CAFO) and aquaculture supported by precipitation fed water stocks and flows [temperature, wind, and precipitation extremes – impacts on crops, animals, facilities; drought and heat wave impacts on water supplies] • Land for forestry [drought, elevated temperatures; wild fires; flooding, erosion] • Water for irrigation and

Bi-Directional, Bilateral Interdependencies
[Examples of Climate Linkage Shown in Brackets in Table]

Applications/Systems/Processes	Energy-Water (E↔W) Interface Linkage Functions f_{EW} (N_{EW}, E_{EW}) and f_{WE} (N_{WE}, E_{WE}) [see Fig 2.3, Fig A.1, Table A.1]	Energy-Land (E↔L) Interface Linkage Functions f_{EL} (N_{EL}, E_{EL}) and f_{LE} (N_{LE}, E_{LE}) [see Fig 2.3, Fig A.1, Table A.1]	Land-Water (L↔W) Interface Linkage Functions F_{LW} (N_{LW}, E_{LW}) and f_{WL} (N_{WL}, E_{WL}) [see Fig 2.3, Fig B.1, Table A.1]
	wastewater during extreme weather and flooding events]	winds; changes in reliability of wind and solar resource]; • Use of forest wastes and trimmings for bioenergy and biofuels [extreme weather event impacts flooding & erosion; drought & wind promoted forest fires] • Use of ag wastes for energy (crop residues, CAFO, dairies [changes in crop residue] • Use of municipal wastes for energy	aquaculture [drought-water shortages; water quality impacts] • Forest, grasslands, wetlands, ecosystems providing watershed for surface water supply capture and flow, and groundwater recharge [drought, elevated temperatures; wild fires, flooding, erosion]
Energy Mineral Extraction (mines & wells), and Related Fuel Processing/Refining, Transport, Storage, and Distribution	• Water for shale gas fracking [Water quality and quantity impacts with expanded production to supply increased power demand during heat waves; flooding and escape of frack wastewater] • Water for oil and gas, oil shale and tar sands processing [Water quality and quantity impacts with expanded production to supply increased power demand during heat waves; impacts of extreme weather events, flooding, infrastructure damage, and escape of contaminated fracking wastewater] • Produced water from oil, gas, coal bed methane (CBM) wells • Energy for treating produced water and	• Energy/power for mine and well energy mineral extraction operations, transport, processing/refining to fuels, storage and distribution • Land use to mine/extract energy resource deposits [extreme weather event impacts on infrastructure; flooding; drought leading to water shortages] • Land use for mining fertilizer minerals for biofuels • Land use for fuel processing and power generation [extreme weather event impacts on infrastructure; flooding; drought leading to water	• Watershed impacts of: -Water for mining -Water from energy production • Water quality impacts of: -Leaching of tailings -Contaminated mine water -Contaminated produced or fracking water [flooding and escape of contaminated wastewater- erosion and transport of contaminated sediment] • Treated produced water used for land reclamation • Land cover vegetation change with impacts on water capture and recharge of surface and

	Bi-Directional, Bilateral Interdependencies [Examples of Climate Linkage Shown in Brackets in Table]		
Applications/Systems/Processes	**Energy-Water (E↔W)** Interface Linkage Functions f_{EW} (N_{EW}, E_{EW}) and f_{WE} (N_{WE}, E_{WE}) *[see Fig 2.3, Fig A.1, Table A.1]*	**Energy-Land (E↔L)** Interface Linkage Functions f_{EL} (N_{EL}, E_{EL}) and f_{LE} (N_{LE}, E_{LE}) *[see Fig 2.3, Fig A.1, Table A.1]*	**Land-Water (L↔W)** Interface Linkage Functions F_{LW} (N_{LW}, E_{LW}) and f_{WL} (N_{WL}, E_{WL}) *[see Fig 2.3, Fig B.1, Table A.1]*
	• fracking wastewater from energy mineral extraction • Water for energy and power production to operate mines, wells, pumps, refineries •	shortages] • Land use for transport (roads, railways, pipelines, power lines), storage (tanks and depots), and distribution of fuels and power [extreme weather event impacts on infrastructure]	ground supplies [drought and hot or cold extreme temperature impacts; flooding and erosion]
Electric Power Generation **Hydroelectric** **Thermoelectric (Fossil, Nuclear, & Biopower),** **Geothermal** **Renewable Solar & Wind** **Transportation (Land, Air, Marine)** **Petroleum fuels** **Biofuels** **Electric & hybrid vehicles** **Municipal, Industrial, Commercial, Residential, and Government** **(Federal/State/Local) Facilities and Operations** **Water Resources Pumping, Treatment, Use Conditioning**	• Water for hydropower [drought – reduced supplies; higher temps – reduced snowpack and earlier snowmelt altering supply flow timing; increased reservoir evaporation loss] • Water for thermal power cooling [fossil, nuclear, biomass]; [drought – reduced supplies; higher temps – reduced snowpack and earlier snowmelt altering supply flow timing; heat waves – increased power demand & stress on water supplies] • Water for geothermal power • Water for solar thermal generation [drought – reduced supplies; higher temps – reduced snowpack and earlier snowmelt altering supply flow timing; heat waves – increased power demand & stress on water supplies] • Reduced water demand from expansion of wind and PV	• Land use for producing biomass for biopower [drought, heat wave, flooding impacts on biomass crop production] • Land for thermal power plants [high temperatures reduce thermal cooling efficiency;] • Land for hydropower [drought in watershed; reduced watershed precipitation & snow pack; flooding & siltation] • Land for geothermal power • Land for transmission lines [extreme weather events – damage to infrastructure] • Land for wind and solar energy [wind and solar resource changes impacts wind & solar generation] • Land for nuclear waste repositories • Electric power for farms, towns,	• Watershed precipitation and snow pack capture & storage for surface waters and groundwater recharge [drought in watershed; reduced watershed precipitation & snow pack; increased evapotranspiration]Surface water storage reservoirs, lakes, streams, rivers • Ground water [prolonged drought ; overpumping] • Ecological interactions with cooling water discharge • Ecological and agricultural interactions with dam release timing and flow variations [drought – reduced supplies; higher temps – reduced snowpack and earlier snowmelt altering supply flow timing; flooding;

Applications/Systems/Processes	Bi-Directional, Bilateral Interdependencies [Examples of Climate Linkage Shown in Brackets in Table]		
	Energy-Water (E↔W) **Interface Linkage Functions** f_{EW} (N_{EW}, E_{EW}) and f_{WE} (N_{WE}, E_{WE}) [see Fig 2.3, Fig A.1, Table A.1]	Energy-Land (E↔L) **Interface Linkage Functions** f_{EL} (N_{EL}, E_{EL}) and f_{LE} (N_{LE}, E_{LE}) [see Fig 2.3, Fig A.1, Table A.1]	Land-Water (L↔W) **Interface Linkage Functions** F_{LW} (N_{LW}, E_{LW}) and f_{WL} (N_{WL}, E_{WL}) [see Fig 2.3, Fig B.1, Table A.1]
	[Reduction in PV performance at elevated temperatures; Impacts of changing insolation and wind conditions; infrastructure damage from extreme weather events] • Power generation for electric and hybrid vehicles and mass transit • Water for transportation of fuels [drought, heat, and flood impacts on biomass production; extreme weather event impacts on transportation fuel production infrastructure; reductions in water supplies – increased water competition for shale gas fracking, etc.] • Energy/power for water pumping, transport, treatment and conditioning for end-use [extreme weather event impacts on infrastructure]	• cities, municipalities, industrial and commercial facilities, residential • CHP for industry operations • Land and infrastructure for fossil power plant CCS [extreme weather events – damage to infrastructure]	nutrient-loaded tail water from crop irrigation
Agriculture, Aquaculture, Forestry **Eco-System Health and Services**	• Water for biofuels/bioenergy feedstock production and conversion processing [drought, prolonged heat wave; leading to crop damage] • Feedback impact on ecosystems of wastewaters from bioenergy production (nutrient loading), produced and fracking water from energy mineral extraction (contamination), and water used for power plant cooling (thermal loading) [algae blooms in nutrient rich water during periods of high temperatures;	• Atmospheric carbon capture and storage (CCS) in land cover vegetation [extreme weather event impacts; flooding/erosion; drought & heat leading to water shortages and plant kill] • Use of agriculture land for siting of renewable energy production (wind, solar, biomass) [extreme weather event impacts on infrastructure; flooding; high winds; changes in reliability of	• Land for farm, ranch, CAFO agriculture [temperature, wind, and precipitation extremes – impacts on crops, animals, facilities] • Land for forestry [drought, elevated temperatures; wild fires; flooding, erosion] • Land for aquaculture [extreme weather event impacts on infrastructure] • Water for irrigation and aquaculture

Applications/Systems/Processes	Bi-Directional, Bilateral Interdependencies [Examples of Climate Linkage Shown in Brackets in Table]		
	Energy-Water (E↔W) **Interface Linkage Functions** f_{EW} (N_{EW}, E_{EW}) and f_{WE} (N_{WE}, E_{WE}) *[see Fig 2.3, Fig A.1, Table A.1]*	**Energy-Land (E↔L)** **Interface Linkage Functions** f_{EL} (N_{EL}, E_{EL}) and f_{LE} (N_{LE}, E_{LE}) *[see Fig 2.3, Fig A.1, Table A.1]*	**Land-Water (L↔W)** **Interface Linkage Functions** F_{LW} (N_{LW}, E_{LW}) and f_{WL} (N_{WL}, E_{WL}) *[see Fig 2.3, Fig B.1, Table A.1]*
	spillage of energy mineral extraction wastewater during extreme weather and flooding events]	wind and solar resource]; • Use of forest wastes and trimmings for bioenergy and biofuels [extreme weather event impacts flooding & erosion; drought & wind promoted forest fires] • Use of ag wastes for energy (crop residues, CAFO, dairies) [changes in crop residue] • Use of municipal wastes for energy	[drought-water shortages; water quality impacts] • Forest and ecosystem based watersheds for surface supply capture and groundwater recharge [drought, elevated temperatures; wild fires, flooding, erosion]
Mining, Energy Mineral Extraction: **and** **Related Fuel Processing/Refining, Transport, Storage, and Distribution**	• Water for shale gas fracking [Water quality and quantity impacts with expanded production to supply increased power demand during heat waves; flooding and escape of frack wastewater] • Water for oil and gas, oil shale and tar sands processing [Water quality and quantity impacts with expanded production to supply increased power demand during heat waves; impacts of extreme weather events, flooding, infrastructure damage, and escape of contaminated fracking wastewater] • Produced water from oil, gas, coal bed methane (CBM) wells • Energy for treating produced water from energy extraction	• Land use to mine/extract energy resource deposits [extreme weather event impacts on infrastructure; flooding; drought leading to water shortages] • Land use for mining fertilizer minerals for biofuels • Land use for fuel processing and power generation [extreme weather event impacts on infrastructure; flooding; drought leading to water shortages] • Land use for transport (roads, railways, pipelines, power lines), storage (tanks and depots), and distribution of fuels and power [extreme weather event impacts	• Watershed impacts of: -Water for mining -Water from energy production • Water quality impacts of: -Leaching of tailings -Contaminated mine water -Contaminated produced water [flooding and escape of contaminated wastewater- erosion and transport of contaminated sediment] • Treated produced water used for reclamation • Land cover vegetation change with impacts on water capture and recharge of surface and ground supplies [drought and hot or cold extreme

Applications/Systems/Processes	Bi-Directional, Bilateral Interdependencies [Examples of Climate Linkage Shown in Brackets in Table]		
	Energy-Water (E↔W)	Energy-Land (E↔L)	Land-Water (L↔W)
	Interface Linkage Functions f_{EW} (N_{EW}, E_{EW}) and f_{WE} (N_{WE}, E_{WE}) *[see Fig 2.3, Fig A.1, Table A.1]*	Interface Linkage Functions f_{EL} (N_{EL}, E_{EL}) and f_{LE} (N_{LE}, E_{LE}) *[see Fig 2.3, Fig A.1, Table A.1]*	Interface Linkage Functions F_{LW} (N_{LW}, E_{LW}) and f_{WL} (N_{WL}, E_{WL}) *[see Fig 2.3, Fig B.1, Table A.1]*
	• Water for energy and power production to operate mines, wells, pumps, refineries • Fuel for vehicles associated with operation of mines, fuel production, transport, and distribution	on infrastructure]	temperature impacts; flooding and erosion]

Distribution

www.ingramcontent.com/pod-product-compliance
Lightning Source LLC
Chambersburg PA
CBHW080640180526
45168CB00008B/3244